P9-APG-320

WITHDRAWN

Nanosciences
The Invisible Revolution

WITHDRAWN

Figure Captions for the Cover Page (Top to Bottom)

"NANO" written with 53 Au atoms (see p. 4 for details).

Bubble 1:

One molecular orbital of a single methylterrylene molecule recorded by C. Villagomez and S. Gauthier from the author's Toulouse CNRS research group using a low temperature UHV-STM Omicron microscope (*Chem. Phys. Lett.* **450**, 107 (2007)).

Bubble 2:

STM image of a single molecular wire Lander molecule (the four yellow protusions arranged in a square) in electronic contact with an atomic copper wire (the small red expansion on the left). Image recorded by L. Grill at the Berlin Frei Universitat Physics Institute STM research group on the lab's low temperature UHV-STM (*Nano Lett.* **5**, 859 (2005)).

Bubble 3:

A surface molecular model of a single-molecule motor well-positioned between its two driving metal pads. The molecular motor was designed and chemically synthesized by G. Rapenne and J. P. Launay from the author's Toulouse CNRS research group.

Bubble 4:

STM image of an ultra-clean Si(111) surface. Each white spot is a single silicon atom. Image recorded by Jianshu Wang on an UHV-STM Omicron microscope at the author's Singapore A*STAR-IMRE Institute research group.

Bubble 5:

STM image of the Hoover molecule, an organic molecule deliberately designed to clean a surface atom by atom by collecting up to six atoms under its organic board. Image recorded by L. Grill at the Berlin Frei Universitat Physics Institute STM research group on the lab's low temperature UHV-STM (*Nature Mater.* **4**, 892 (2005)).

The cartoon of a serious scientist holding a single molecule was drawn by the artist Midam (M. Ledent) in the Kid Paddle journal (Ed. Dupuis, Belgium). Kid Paddle © Dupuis, 2008 by Midam. www.dupuis.com.

MONTGOMERY COLLEGE
ROCKVILLE CAMPUS LIBRARY
ROCKVILLE, MARYLAND

Nanosciences

The
Invisible
Revolution

Christian Joachim
CEMES-CNRS, France

Laurence Plévert
Independent Science Journalist

Translated by
John Crisp

 **World Scientific**

NEW JERSEY · LONDON · SINGAPORE · BEIJING · SHANGHAI · HONG KONG · TAIPEI · CHENNAI

1287018

JAN 2 1 2011

Published by

World Scientific Publishing Co. Pte. Ltd.

5 Toh Tuck Link, Singapore 596224

USA office: 27 Warren Street, Suite 401-402, Hackensack, NJ 07601

UK office: 57 Shelton Street, Covent Garden, London WC2H 9HE

British Library Cataloguing-in-Publication Data
A catalogue record for this book is available from the British Library.

NANOSCIENCES: THE INVISIBLE REVOLUTION

Copyright © Editions du Seuil, 2008

Translation Copyright © 2009 by World Scientific Publishing Co. Pte. Ltd.

All rights reserved. This book, or parts thereof, may not be reproduced in any form or by any means, electronic or mechanical, including photocopying, recording or any information storage and retrieval system now known or to be invented, without written permission from the Publisher.

For photocopying of material in this volume, please pay a copying fee through the Copyright Clearance Center, Inc., 222 Rosewood Drive, Danvers, MA 01923, USA. In this case permission to photocopy is not required from the publisher.

ISBN-13 978-981-283-714-1 (pbk)
ISBN-10 981-283-714-0 (pbk)

Typeset by Stallion Press
Email: enquiries@stallionpress.com

Printed in Singapore by Mainland Press Pte Ltd.

ACKNOWLEDGMENTS

Thanks are due to Karim Benzerara, Philippe Bertran, Christophe Bonnin, Éric Finot, Robert Folk, Henk Kubbinga and Luc Montagnier; to the archives department of the French Academy of Sciences; to the A*STAR VIP Atom Technology group at IMRE in Singapore and to the "Nanosciences" group at CEMES/CNRS in Toulouse, France.

CONTENTS

INTRODUCTION

INFINITIES IN A GRAIN OF SAND

This book is an invitation to take the plunge into the infinitely small, to stay there — "at the bottom" — and play with a single atom or a single molecule. In our day-to-day lives we never encounter such objects individually, because they are too small to be grasped one by one. So how big is an atom? Rather than talking about its dimensions — one ten-billionth of a meter, or ten millionths of a millimeter — another approach is to consider how big it would be if you, the reader of this book, were the size of the Earth. In that case, an atom would be a tiny ball one millimeter across. And if the atom were as big as a room, you would be tall enough to touch the Sun. So an atom is invisible not only to the naked eye, but even to our most powerful optical microscopes.

In 1981, a new microscope was invented: the so-called scanning tunneling microscope (STM), which can provide an image of a single atom or a single molecule on a computer screen. However, as far back as the 1950s, "electron" microscopes had shown images of atoms on a phosphorescent screen. What is different about the STM is that its minute tip can now also touch a single atom at a time, and move it at will. Usually, when we touch something, the billions of atoms in our fingers come into "contact" with the billions of atoms in the object. But the tip of the STM is so sharp that it allows us to touch a single atom and even to assemble new atomic architectures atom by atom. The tip becomes an extension of the scientist's or engineer's finger.

The result is that the STM is revolutionizing our relationship with matter. With this microscope as a tool, a different technological method has emerged, in which ever-larger edifices are constructed atom by atom and "monumentalized" until each embodies a minute but functioning machine. It is a bottom-up approach to machine construction, the reverse of miniaturization.

Suppose, for instance, that we wanted to make a cube a million times smaller than a grain of sand, with edges one nanometer — i.e. one billionth of a meter — long. To build this nanocube, we would have to put together around sixty atoms one by one. This can be done with the STM, and this bottom-up technology of atom-by-atom construction is called "nanotechnology." In the top-down approach of miniaturization, we would have to remove 100 billion billion atoms to make the same nanocube from a cube that started out with edges one centimeter long.

In essence, therefore, nanotechnology is a technology that is sparing with material resources. Over the years, however, the definition has become more elastic. Nanotechnology has become "nanotechnologies," which are no longer just concerned with the atom-by-atom manipulation of matter, but also encompass all the techniques used to make "small objects" with a precision measured in nanometers, even though they bring billions of atoms — rather than just a few — into play.

How did we get from nanotechnology, with its focus on sustainable development, to the "catch-all" nanotechnologies we see today? This shift, which we will describe in the first chapter, is the result of complex political maneuverings involving vested interests, money and competition. Within the space of a few years, nanotechnology was diverted from its initial purpose. Today's nanotechnologies, which employ the same technological principles as before the invention of the STM, push miniaturization to its limits and flirt with the nanometric scale. They have produced extraordinary devices a few tens of nanometers in size, but small as they are, these devices still contain several thousand atoms. We will describe these minimarvels in Chapter 2, which will retrace the key episodes in the adventure of miniaturization, a process often wrongly equated with nanotechnology.

In Chapter 3, we tell the true story of nanotechnology. Physicists have long dreamed of working with single atoms or molecules. With the STM, that dream has become a reality. They can now access a single molecule and study it as if by touch. The exploration of this material world at the bottom is only in its infancy. Physicists want to know whether the phenomena observed at this scale obey the laws we know or will force us to rethink our understanding of physics.

With the STM, it is in principle possible to build every kind of possible or conceivable molecular structure atom by atom. True, there is nothing new about synthesizing new molecules. Hundreds are produced every day in laboratories for use in colorings or drugs. However, these new molecules are manufactured by the billion (a single drop of water contains more than 1500 billion billion molecules!), whereas the molecule constructed at the tip of the STM is a single entity.

Equipped with the STM, physicists and chemists have an opportunity to design new molecules, to produce the tiniest of nanomachines, such as mechanical devices or computers. This bottom-up approach is promising: the workings of fantastic little machines made up of a single molecule have already been studied, and in Chapter 4 we will see what the next generations of such machines may look like.

One day, it will probably be possible to assemble larger structures, for example molecules from the living world such as DNA, together with proteins and membranes to enfold them. However, once this molecular apparatus has been assembled, what will happen? Does life lie at the end of monumentalization? Our fascination with life belongs to the sphere of the sacred. Where, in the monumental mass of atoms that constitute a cell, is the essence of life to be found? In Chapter 5, we will consider whether the fantasy of recreating life from nonliving matter has any chance of becoming a reality.

Nanotechnologies raise other disconcerting questions. Is there a risk that nanomachines could break free from our control? Might they be poisonous or damaging to the environment? Nanotechnologies arouse heated debate, but beyond that they raise questions about scientific progress, about the balance between the benefits and the risks of their applications. In Chapter 6, we will summarize these debates and consider whether there is anything to fear in the exploitation of the infinitely small.

The purpose of this book is to describe what nanotechnologies really are, and to consider their scientific and technical consequences. To do this, we need to rediscover the urge to know, which is such a characteristic feature of the human mind. We are more accustomed to directing that questing spirit toward stars and galaxies, toward the immensely large. But there are infinities too in a grain of sand.

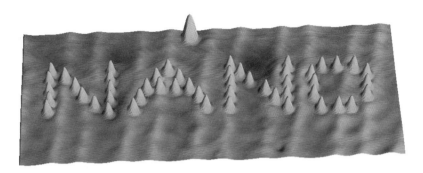

Scanning-tunneling-microscope image of 51 gold atoms (plus one unidentified atom) deposited on the surface of a gold crystal. In this picture, the atoms look like small bumps 0.15 nanometers high (nanometer: one millionth of a millimeter). Each of the gold atoms has been moved by the tip of the microscope to write "NANO." The positioning precision is 0.05 nanometers and the smallest distance between two atoms of gold in the letters is 1.2 nanometers. This atom-by-atom image was produced for this book on a Createc low temperature microscope by Soe We-Hyo and Carlos Manzano, researchers in the "Atom Technology" group at the IMRE Institute of Singapore's A*STAR agency, headed by the author. Real size of the image: 10 nanometers × 30 nanometers.

CHAPTER 1

A CASE OF MISDIRECTION

In the 1980s, nanotechnology offered the potential fulfillment of a dream for anyone concerned with the future of the planet. It was becoming apparent that we would one day need to reduce the quantity of materials and energy used in making all our machines. Today, for example, manufacturing a PC consumes 240 kilograms of fossil energy, 22 chemical products and 1500 liters of water. Producing a single USB key entails 250 liters of water and numerous chemical pollutants.[1]

Our hope was that nanotechnology, then in its infancy, would liberate industry from the mass use of materials, and usher in a new era of sustainable development. That was the aspiration I shared with a number of other researchers. At the time, I drew up my own state-of-the-planet assessment, which still reminds me of that youthful aspiration. I used to amuse myself by looking for new, environmentally safe technologies in different industrial sectors. One of these was the research by Kevin Ulmer, research director at Genex in California, on the possibility of manufacturing ultraminiaturized electronic circuits using proteins produced by genetically programmed bacteria. His ideas were only one step away from the edible computer! The work of the American chemist Ari Aviram, at IBM's T. J. Watson research labs near New York, was also high up on my list. His aim was to design a single molecule that would act as an electronic component. He was working on an electrical switch consisting of a current-blocking molecule, or a molecular electrical rectifier — in other words, a molecule that would only allow current to flow in one direction.

---

[1] Ruediger Kuehr and Eric Williams, *Computers and the Environment: Understanding and Managing their Impacts* (report for UNESCO) (Kluwer, Dordrecht, 2003).

In the same vein, Forrest Carter, a US chemist working at the NRL (Naval Research Laboratory), imagined making not just single components, but whole electronic circuits, from just a few molecules. These projects raised the prospect that microelectronics might become "nanoelectronics," mitigating the environmental impact of the electronics industry.

I was not the only one looking for more environmentally friendly alternative technologies. In his doctoral thesis, Eric Drexler, a young engineer at MIT (Massachusetts Institute of Technology) in Boston, had imagined devices other than molecular components and electronic circuits. In his 1986 book *Engines of Creation*,[2] he describes a very remote future when molecular machines would recycle waste and produce pure water and energy. These machines, stripped to their essentials, i.e. to a few molecules, would take our civilization into the era of molecular technology.

However, these fine projects were absorbed by nanotechnologies in the broad sense — those based on the classic techniques of micromanufacture — and finally came to nothing. Today, people do not associate nanotechnologies with hopes for less resource-hungry industries, but with anxieties of different kinds: Might they be poisonous? Might they escape our control? We will return to these questions in Chapter 6. How did we get to this point? What caused the shift from all that environmental promise to this sense of mistrust? An analysis of the emergence of what has come to be called the "nano bubble" casts light on this change in the values associated with nanotechnology.

Political Hijacking

It was a time of promise, the start of the nanotechnology adventure, the 1980s. A few researchers, myself included, were working on molecules capable of performing electronic functions and were moving into the field of molecular electronics opened up by Ari Aviram and Mark Ratner. Others were exploring a fantastic new instrument, the scanning tunneling microscope, invented in 1981 by G. Binnig and H. Rohrer (Nobel Laureates in 1986), which was capable of "seeing" atoms and molecules and, above all, of manipulating them individually, as was invented by D. Eigler

---

[2] *Engines of Creation: The Coming Era of Nanotechnology* (Anchor, New York, 1986).

in 1989. They launched the first experiments to be conducted directly on a single molecule. However, these experiments made little impact and the scientific fraternity — especially in Europe — was often skeptical as to the potential of the STM. Nanotechnology was still a small field. By 1995, there were only five teams capable of atomic scale manipulation: three in the USA, one in Europe and one in Japan. It was my good fortune, having worked with Aviram, to be a member of one of these pioneering teams, under Jim Gimzewski at the IBM laboratory in Zürich, which was attempting to manipulate ever-larger molecules with an STM. Few though we were, we made progress and discovered, sometimes by chance, some curious phenomena, which will be described in Chapters 3 and 4. Research should have continued along these lines, but that is not what happened.

Instead of that, events in the mid-1990s took a different turn, driven not by scientific research, but by politics. It all began in the USA, where pressure groups prevailed upon Congress and the Clinton administration to launch a major program called the National Nanotechnology Initiative (NNI). It is worth dissecting the origins of this program to understand how and why nanotechnology diverged from its original calling (the manipulation of atoms) and its original purpose (ecotechnology) to follow a quite different path in the NNI, to become "nanotechnologies" and to be swallowed up by the global technosphere, first in the USA and then worldwide.

In June 1992, ecotechnology was all the rage. Tennessee senator Albert Gore came back from the second Earth Summit in Rio de Janeiro with his environmental instincts so fired up that he organized a Senate hearing for the top US specialists on the topic of "new technologies for sustainable development." One of the people heard was Eric Drexler, whose book had aroused a surge of interest. When it was published in 1986, the manipulation of atoms to produce nanocomputers and other nanomachines was no more than a very hypothetical possibility, disputed by many scientists. In 1989, however, the world discovered that the STM could be used to move atoms around. Suddenly, Drexler and his book had gained new credibility. On June 26, 1992, therefore, he was invited to address the committee of American senators assembled by Al Gore. His talk was remarkably unsensational. He explained that building a machine molecule by molecule could be a cleaner and more efficient process than was possible with existing technologies. To lend the project scientific credibility, he cleverly cited the name of Richard Feynman, winner of the Nobel Prize for Physics, unearthing a 1959 speech in which the illustrious physicist had mentioned nanometric scale manufacture. And, finally, he played the hole cards of national pride and competitive spirit, pointing out, quite correctly, that the Japanese were investing heavily in research on the manipulation of atoms.

According to the minutes of the hearing, Al Gore was won over to the project described by Drexler in the Q&A session. Although he had started out simply interested in miniaturization, within a few minutes he signed up for this nanotechnology, with its potential to assemble "molecule machines" directly from atoms and molecules. A technological development pressure group was then created, with Drexler playing the role of a modern Jules Verne, a master of technological prediction, and founding his California-based Foresight Institute.

During this time, as Drexler had mentioned, the Japanese government had launched a research program on the manipulation of atoms. Its goal was to support the future of the country's microelectronics sector and not to fall behind the Americans! Results were not slow in coming. At Japan's Riken public research laboratory, a researcher called Masakura Aono was working on the construction of atomic scale circuits. He succeeded in developing atom-by-atom engraving on a silicon surface using an STM. The Japanese and the Americans were now neck and neck.

Back to politics. Bill Clinton, elected president in 1992, put his Vice President, Al Gore, in charge of new technologies. The USA was facing a number of challenges. The end of the Cold War had changed US research priorities, shifting the focus to international competitiveness. The goal was no longer simply to support military research, but to strengthen R&D programs on nonmilitary consumer goods. The excellent health of the Japanese and Korean electronics industries was giving US industrial bosses sleepless nights. To protect American research, it was essential to re-resource the universities, which were working with largely obsolete equipment. Gore was essentially given the task of reorganizing US research, a job that involved money, lots of money. Drawing inspiration from the great scientific program headed by President Roosevelt at the end of World War II, he delivered his conclusions in August 1994, in a report entitled *"Science in the National Interest."*

Does the report promote the ecotechnology project that had so excited Gore two years earlier? Alas, it contains little if anything on sustainable development .... The Vice President's initial interest in the environment had evaporated along the way.[3] Instead of supporting research on the manipulation of atoms and molecules, which might foster a more environmentally friendly industry in the future, the report proclaims that nanotechnologies are strategic to America's current industrial development. Nanotechnologies have suddenly become crucial, not to the sustainable development of the planet, but to the immediate future of the national microelectronics, chemicals and pharmaceuticals industries. So what had happened in those few years?

Given the scale of the challenge of reorganizing US research — and the money that went with it — an industrial lobby had stepped up to influence the contents of the report. The "pro-sustainable-development" pressure group that had gravitated toward Al Gore, with Eric Drexler as a key player, had to give way to this new rival. In the space of two years, Drexler's star had waned. He was under attack from many scientists critical — not without reason — of the lack of a scientific basis in his work. Some American newspapers even began to liken him to the guru of a cult, his Foresight Institute. Gradually, he lost influence and credibility.

The industrial lobby found its champion in the person of Mihail Roco. A former professor of mechanical engineering, Roco was appointed in 1990 to head the Engineering division of the US National Science Foundation (NSF). In 1995, he launched a research program on the use of nanoparticles in materials. For this program, he requested and got the green light from the Director of the NSF, Neal Lane, Professor at Rice University in Texas, who in 1998 would become scientific adviser to President Bill Clinton. Roco was a reasonable and tenacious academic who for five years had fought hard to build American nanotechnology. For him, Eric Drexler's ideas of molecular factories were fantasies, and nanotechnology — in the sense of molecular technology — had little future. He was one of those who were convinced that the top-down technological approach of miniaturization was the only valid option. He considered the term "nanotechnologies" to encompass all the technologies of miniaturization that operated within spitting distance of the nanometric scale.

---

[3]His interest in sustainable development was to re-emerge in 2006, with the film *An Inconvenient Truth*, and would win him the Nobel Prize for Peace in 2007.

In early 1997, Tom Kalil, President Clinton's economic adviser, contacted Roco. He had read Al Gore's report and wanted to assess the possible economic implications of the nanotechnologies. With the help of Kalil, Roco then set up a working group which, after two years, culminated in the creation of the NNI. They had to convince a dozen agencies responsible for funding nanotechnologies in the USA, draft a plan that could command a consensus and above all deal with opposition from senators who wanted to see funding go to other programs.

There was reluctance in some quarters about the idea of raiding the nation's coffers for these nanotechnologies, however strategic they were supposed to be. To persuade these reluctant senators and dangle the carrot of a molecular technology radically different from anything that had come before, Neal Lane turned to Drexler's book. Roco, however, continued to make sure that the NNI would remain untainted by Drexler's molecular machines. The industrial lobby did its job well: following a final meeting attended by Roco, Lane and Kalil, the NNI was launched on March 11, 1999.

Its initial budget of US$300 million for the year 2000 was a matter of concern to certain senators. With so little money, they feared, US scientific discoveries in nanotechnology could be exploited by Japan and Europe, which might develop new technologies faster than the USA. Competition was intense: no way were they going to be beaten to the draw! In the end, the NNI's budget grew over the years to US$970 million in 2005. The NNI has survived the political changes in Washington. The 2008 budget approved by President George W. Bush was for US$1.447 billion. There is no question that removing sustainable development from the NNI agenda was a boon for the survival of the program, given President Bush's attitude to that concept ....

The Temporary End of Sustainable Industrial Development

On June 22, 1999, a short time after the launch of the NNI, the US House of Representatives organized a new hearing on the NNI's finances. One of the key speakers was Richard Smalley, Professor of Chemistry at Rice University in Texas. Wrapped in the prestige of the Nobel Prize for Chemistry, which he had received in 1996 for the discovery of fullerenes, he was to play such an important role that a third pressure group formed around him. To defend the NNI cause, this Nobel chemist transformed himself into a Nobel nanotechnologist and became the spokesman for a section of the chemistry fraternity. He was smart. In all his appearances before the House of Representatives, between June 1999 and April 2002, he was careful to choose big problems that would strike a chord with the public, such as cancer or energy resources, and to link them with nanomaterials, and hence with nanotechnologies. Exploiting his Nobel prestige, he continued the trend begun by Mihail Roco, of passing a large proportion of materials science off as nanotechnology. Through this approach, he managed to make sure that anybody in the USA doing research in chemistry and materials science got a share of nanotechnology funding.

So, in the end, via a succession of hearings, committees and programs, the NNI's scientific base lost touch with Al Gore's shiny original project for sustainable development. The NNI now covers all of materials science without distinction — from microelectronics to new fuels to biotechnology. It has redefined nanotechnologies so broadly that they now span a wide range of very different techniques and domains. With Eric Drexler out of the picture, Richard Smalley published a series of press articles in 2003 asking him to put an end to his molecular fantasies.

The Planet Goes Nano

On January 21, 2000, President Bill Clinton officially announced the creation of the NNI program at the California University of Technology. A highly symbolic location. Here it was, in 1959, that Richard Feynman had given the speech cited by Eric Drexler in his appearance before the Senate committee chaired by Al Gore. Later, this speech would come to be considered — quite erroneously — as the starting point of nanotechnology. With this political hijacking of Feynman's scientific prestige, the wheel had come full circle.

The NNI's scientific rankings were now headed by the big guns of microelectronics, materials science and biotechnology (rechristened nanobiotechnology for the occasion). Atomic manipulation, molecular electronics and the first prototype molecule machines were relegated to the bottom of the pile.

No country would resist this US definition of nanotechnology. The NNI was the symbol of America's resurgence, and alarm bells were ringing on every continent: "What if the Americans pull off another 'first man on the Moon' coup, but this time planting the flag of technology in the infinitely small?" The European Commission in Brussels and every nation in Europe began scrabbling through their archives on the off chance that there might have been a few projects in the 1990s with a hint of "nanotechnology" as defined by the NNI, just as a face-saving move. Of course, they found such projects in materials and microelectronics, in the race to miniaturize electronic chips. Europe's honor was saved. This would be the new focus of research.

The streetsmart operators of research in Europe — and many other parts of the world — grasped the opportunity to get funding for their activities. They jumped aboard the NNI bandwagon, without asking themselves what nanotechnologies really were — what was to be gained? If a European microelectronics or microtechnology lab was getting antsy about future funding, all it had to do was call itself the "European Nanotechnology Centre" to resolve the problem. If a chemistry lab in Germany, Switzerland or France was on the slide, the prefix "nano" would quickly put it back in the frame. If a materials science laboratory needed new hardware, it could raise cash by submitting a "nano-research" project.

In France, a group of experts brought together by the French Observatory for Advanced Technologies (OFTA) worked between 1999 and 2002 to redefine nanotechnology, looking for a definition of the nanotechnology project without reference to the microelectronics and materials science lobbies. But it was already too late. The effect was the same as in the USA. A whole scientific community came into being (if you want funding, there is strength in numbers), calling itself "nano" and effectively defining a new scientific field. A strange way of establishing definitions! Successive French nanotechnology programs were modeled on the big American themes: "miniaturization" for the microelectronics industry, "nanomaterials" for the chemistry fraternity, and a new "biotech" sector. The same process has taken place all over the world.

The European Commission went down the same track with the launch of its big NMP (Nanotechnologies, Materials and Processes) program in 2002. This program covers the whole field of materials, and has nothing whatever to say about the manipulation of atoms and molecules. However, it does mention sustainable development and recognizes the possibility that nanotechnology could one day become an ecotechnology. Since the early 1990s, microelectronics has also had its own big program, called IST (Information Society Technologies). The main focus is on the "miniaturization" trend in the electronics industry, but it pays lip service to long-term research, some of it reflecting the "monumentalization" approach that will be explored later in this book. At the end of 2006, IST was replaced by a new, ICT (Information and Communication Technologies) program, which for the first time in a European Commission program mentions atom manipulation and the use of individual molecules to perform an electronic function.

What can we learn from this story? That economic competition and vested interests are often more powerful than scientific aspirations, which often rest on somewhat utopian foundations. The scientific establishments of the different countries arrived at a highly elastic definition of nanotechnology in order to protect and justify this political hijacking. "Nanotechnologies are the design, characterization, production and application of structures, devices and systems by controlling shape and size at nanometer scale," states the UK's Royal Society and Royal Academy of Engineering. According to other definitions, nanotechnologies begin when new physical phenomena appear in samples measuring less than 100 nanometers in at least one direction.

That is the way with scientific progress. Adding to the storehouse of human knowledge takes financial resources, and there is no guarantee of a return on investment, whether technical or cultural (an increase in scientific knowledge). Hence the impression of a technoscientific project that has proceeded nonrationally, almost by alchemy, in order to satisfy all the parties involved.

CHAPTER 2

THE INCREDIBLE SHRINKING CHIP

Some say that the story begins on December 29, 1959. That evening, Richard Feynman, who was to receive the Nobel Prize for Physics in 1965 for his work in theoretical physics, gave an after-dinner speech to a gathering of the elite of American physics. At the age of 41, he had already acquired a huge reputation and was universally seen as the possessor of a remarkable, creative and unconventional scientific intelligence. Once again, he was to spring a surprise on his audience. Like a preacher from the pulpit, he began his speech: "There's plenty of room at the bottom."[4]

A Mythical Speech

Today, this speech is taken as gospel and perceived, erroneously, as seminal, whilst Feynman himself is seen as the father of nanotechnology. The story goes that, inspired by the master's words of wisdom, physicists launched into the exploration of the world at the bottom, thereby creating the field of nano-technology. The reality is different. Although the speech does predate the success of the nanotechnologies, it had absolutely nothing to do with their emergence. Despite all the prestige Feynman already enjoyed, his words that evening aroused little enthusiasm. Paul Shlichta, an American physicist present at the dinner, reports: "The general reaction was amusement. Most of the audience thought he was trying to be funny .... It simply took everybody completely by surprise."[5] In the years that followed, the impact of the

---

[4]Speech delivered at the annual meeting of the American Physical Society at Caltech (California Institute of Technology), and published under the title "There's plenty of room at the bottom: an invitation to enter a new field of physics," in *Engineering and Science*, February 22, 1960.

[5]Chris Toumey, "Apostolic succession," *Engineering and Science*, Vol. 68, No. 1/2, 2005.

speech became no greater, and it slipped into oblivion. The European physicists Gerd Binnig and Heinrich Rohrer, inventors of the scanning tunneling microscope, and the American physicist Don Eigler, the first man to have written on the surface of a metal with individual atoms, have been asked how this speech influenced their work. Their answer is unambiguous: the speech had no influence, for the very good reason that they had never heard of it. Feynman, who died in 1988, witnessed the emergence of scanning tunneling microscopy and the advances in microelectronic and micromechanical miniaturization. However, he made no claims of having fathered these achievements. Nor did he ever make any connection between these advances and his 1959 speech. He often discussed the physics of computers in his classes at Caltech, but he never worked on problems linked in any way, however remotely, with nanotechnologies. On only one other occasion in his life did he refer to the themes of "There's plenty of room at the bottom," in an article published in 1983. Feynman's speech only became famous in the 1990s, when Eric Drexler unearthed to give credence to his own ideas.

So what did Feynman actually say that evening? "I would like to describe a field in which little has been done, but in which an enormous amount can be done in principle … it would have enormous technical applications. … What I want to talk about is the problem of manipulating and controlling things on a small scale. … It is a staggeringly small world that is below."

He deserves credit for recognizing the importance of miniaturization and pointing out the need to explore the world at the bottom. Does that make him a visionary in this field? He certainly asked himself: "What would happen if we could arrange the atoms one by one the way we want them?" But he came up with no solutions and devised no instruments capable of doing it. He refers to ultimate precision in manufacture, not the ultimate size of the devices themselves. The speech is an exercise in futurology. He points out how often physicists achieve results when they try to push the envelope. He cites the example of two physicists — one who sought to achieve ever-lower temperatures, the other ever-higher pressures — who in so doing opened up new fields of research. Why not, asks Feynman, push the miniaturization of devices and machines to the limits? He raises the possibility of storing information with a few hundred atoms. But we now know that a single atom could be enough.

Feynman did not, as is often claimed, foresee the emergence of nano-technology. Nor was he the first to raise the question of the limits of miniaturization and the exploration of the very small, of the world at the bottom, as a glance back shows. "As a young man, it was in my mind to become an inventor, a Newton ... the world of details remains to be explored; it is another world, the most important of all, that I believed myself to have discovered," confessed Napoleon to the mathematician Gaspard Monge, on board the frigate *La Muiron*, returning from the Egypt expedition in 1799. Feynman was a peerless physicist, but he is one of a long line of physicists — and nonphysicists — who have taken an interest in the world at the bottom and in miniaturization, and who have even ventured there.

The Giants of Miniaturization

When can we place the beginnings of miniaturization? Greek scientists built magnificent astronomical clocks, with mechanisms made from small cogwheels. These miniature representations of the solar system are technological marvels. Later, advances in clockwork would play an essential role in the miniaturization of the mechanisms used in automata and then in robots. However, the miniaturization of machines is not solely about technical advances. It is inseparable from and has acted as a driving force in the progress of science as a whole.

One day in 1764, John Anderson, a physics professor at the University of Glasgow in Scotland, wanted to explain to his students the workings of the "fire pumps" or "atmospheric machines" used in Britain to pump water out of coalmines. These machines were too big to bring into a classroom, so he had a small version built, less than a meter high. When, to his embarrassment, the miniature machine failed to work, he turned for help to a scientific instrument repair shop in which one of the personnel went by the name of James Watt. The latter identified the cause of the problem: in the miniature version, the atmospheric pressure was insufficient to overcome the friction between the piston and the chamber wall. Watt came up with the idea of using steam pressure instead of atmospheric pressure to lower the piston, which led to the invention of the steam engine, which in turn opened the way to motorized locomotion, since the engine was small enough to fit on a wheeled vehicle. This ushered in a new scientific era, a time of great progress in the science of heat and the development of thermodynamics.

In the course of their research, physicists often encounter problems of measurement, which require instruments of greater and greater precision. In certain cases, miniaturization offers a solution. James Prescott Joule (1818–1889), for example, was looking for a way to measure a minute temperature rise in a bucket of water. Based in Manchester, England, a brewer by trade like his father, he spent his leisure time studying the problem of the equivalence between work and heat. It was known that work could be converted into heat — rubbing two objects together produced heat — and, conversely, that heat could be converted into work (an effect exploited by Watt in his steam engine). Joule wanted to calculate the quantity of heat generated by a given amount of work. He set up an experiment in which falling weights were used to drive blades immersed in a bucket of water, which turned and warmed the water. However, the weights had to be raised forty times an hour to heat the water by a mere half degree. In order to measure this slight rise in temperature, Joule needed a more accurate thermometer than was then available, so he made a miniature thermometer which would be ultraprecise because of its size. It worked like any other thermometer: the expansion of the alcohol (or mercury) in a glass tube is proportional to the rise in temperature, so once the expansion has been gauged and the tube calibrated, the level of alcohol in the tube shows the temperature. Joule made a very fine glass tube and filled it with alcohol, but found that its diameter was not constant. This meant that the expansion of the alcohol in the tube was not proportional to the temperature, which falsified the measurements. Joule identified these irregularities by sliding an optical microscope along the tube and adjusting the gap between the calibrations to compensate for them.

The technique he developed for engraving the calibrations on the tube was ingenious: he coated the tube with beeswax, then made the calibration markings with an extremely sharp knife. He then immersed the tube in an acid solution. The acid attacked the glass where it was not protected by the wax. When he removed the wax, the tube had become an ultraprecise thermometer, with calibrations spaced to a precision of 6 micrometers. Joule had invented the technique of masking and engraving still used in microelectronics today. In 1850, through a combination of this technique and his own perseverance, he was the first to measure the equivalence between work and heat.

From Electron to Electronics

Phenomena in nature are not always easy to study. They are sometimes erratic or hard to reach. To overcome these problems, scientists try to reproduce them in their laboratories, where they can be examined at leisure. In practice, this usually involves the "miniaturization" of the phenomenon. Let us take an example that was to result in the emergence of particle physics, and eventually lead to electronics and the development of microelectronics.

In the 18th century, the French physicist Abbé Nollet and the American Benjamin Franklin were conducting research on lightning and looking for a way to enclose it in some kind of "box" so that they could study it in their laboratories or stage drawing room experiments. It is no secret that studying lightning in nature is not without risk, and that some of their colleagues had received fatal lightning strikes when conducting experiments in thunderstorms. In 1857, Heinrich Geissler, a German manufacturer of scientific instruments, resolved the problem by reproducing real miniature lightning between two electrodes placed in a glass tube filled with gas. In 1874, the English physicist William Crookes perfected this tube for the study of miniature lightning even without gas. The physicists of the time were interested in the nature of the lightning produced in a Crookes tube: was it electromagnetic radiation, as proposed by German physicists, or particles, as upheld by English physicists? It was the Briton Joseph John Thomson who found the answer — in 1898, using a slightly modified version of the Crookes tube, he discovered the electron.

The discovery of the electron ushered in the 20th century era of electronics, which, through unprecedented development in miniaturization techniques, was to culminate in microelectronics. Once again, the starting point was a practical problem. Early telephones were entirely manual: in a telephone exchange, operators had to insert and remove electrical connectors to connect two speakers. However, with the number of telephone subscribers exploding in all the world's great cities, from Paris to New York, the only solution to growing demand was to automate interconnections in telephone exchanges. But how? Engineers began by using small electromechanical relays, followed by vacuum tubes, the famous diodes, triodes and pentodes in our great-grandparents' wireless sets. These descendants of the Crookes tube could function as electrically controlled switches. However, they were big and gave off a lot of heat. This was when the idea arose of replacing the glass tubes by something solid and, if possible, smaller. Three men — John Bardeen, Walter Brattain and William Shockley — all at the laboratories of the Bell Telephone Company in the USA, began work on the problem in the early 1940s. In December 1947, they invented a device based on a small semiconducting crystal, which they called the transistor. Like a triode, a transistor could amplify an electrical signal. But it was smaller and, above all, did not need to be warmed up before it could work.

These transistors were to become the basic components of electronics. Combined, they constitute the logical functions and memories used in all electronic circuits. Initially, however, they had to be etched into the semiconductor crystal one by one and then wired together individually, by manual soldering. This was a fiddly operation carried out under an optical microscope, which generated multiple defects. In 1958, Jack Kilby, an engineer at Texas Instruments, resolved the problem by inventing the first integrated circuit. It was called "integrated" because several electrical components with their interconnecting wires were integrated on the surface of a single material. As a matter of interest, Kilby's colleagues wanted to stack the transistors vertically, whilst Kilby showed that each component could be laid out flat on the surface of a single material. Still by manual soldering, he managed to assemble a transistor, then three resistors and a capacitor, all connected by fine gold wires, on small individual wafers of germanium. A few months later, Robert Noyce from the firm Fairchild Semiconductor succeeded in combining all these components on the surface of a single small wafer of silicon. There was no longer any need for connecting wires between the components, since they were assembled on the surface of the silicon wafer. It was the birth of the integrated circuit (the silicon "chip"). A few months later, mass production of these silicon circuits began.

Enter Gordon Moore

The technology of the integrated circuit was developed as a response to the need to find a way to automate the assembly of electronic components on a single material. So, initially, miniaturization was not the issue. However, it quickly came to the fore in such applications as on-board missile electronics, the systems that control the stability of missiles in flight, using a gyroscope to measure angles and mechanisms to regulate the ejection of combustion gases. Electronic engineers, allied with an army of physicists, used the *Apollo* space program as a springboard for progress in the miniaturization of electronic devices. In missiles and rockets, the savings in volume and weight are a major benefit. However, smaller transistors are also faster and therefore more efficient. The more of them there are on a single chip, the greater the circuit's processing capacity. As a result, transistor density on chips has grown continuously since the 1960s, in line with an empirical law formulated by Gordon Moore in 1965.

Gordon Moore was a young University of California graduate who had worked with transistor coinventor William Shockley, before joining Fairchild Semiconductor. In April 1965, the editor-in-chief of the US magazine *Electronics* asked him to write an article on the future of electronics.[6] At the time, the most complex integrated circuit contained around thirty components, including a few transistors. It was not a lot, but Moore believed in the technology. He noted that, since the invention of the integrated circuit, the number of components had grown from four to eight, then to sixteen the next year, etc.; in other words, more or less doubling every year. The fact is that he had no intention of establishing a law, and simply wanted to put across the message that the components were going to get smaller and the circuits more complex and less expensive. In reality, a fantastic race to miniaturize was about to begin, but he did not know that.

Caltech professor Carver Mead coined the term "Moore's law" for what was initially just an empirical observation. In 1975, Moore readjusted "his" law, when he observed that the density of components on the surface of an integrated circuit was doubling every two years. In the meantime, he had joined forces with Robert Noyce, one of the inventors of the integrated circuit, and in July 1968 they had together created Intel Corporation, which in 1971 produced the first microprocessor, a chip containing 2250

---

[6]Gordon E. Moore, "Cramming more components onto integrated circuits," *Electronics*, April 19, 1965.

transistors.[7] Since then, Moore's law has accurately predicted the growth in the number of transistors on a single chip, which in 2007 had reached the astonishing figure of more than 250 million.

Not only were the transistors smaller, but they worked better. Essentially, halving the size doubles the speed, as the electrons have less distance to travel. At the same time, the power dissipated per transistor diminishes: with four times as many transistors on a chip, the total power lost stays the same, but computational power increases eightfold. In the last 40 years, the size has not just halved as in the example above, but fallen more than a hundredfold! The consequences are obvious: the first computer weighed 50 tons and required 25 kW to process just a hundred or so instructions per second, whereas any microprocessor today weighs only a few grams, can process 100 million instructions per second and consumes a thousand times less energy.

So much progress and so many innovations in so little time! Yet these innovations have changed neither the principles on which transistors work, nor those involved in the manufacture of integrated circuits. Today's transistors are still produced using a masking and engraving process, called lithography, the same method used by Joule to calibrate his thermometer. However, light has replaced the knife he wielded to mark the wax. In optical lithography (or photolithography), the light passes through a mask and insolates a photosensitive resin deposited on a silicon wafer. The mask acts as a stencil, and the light imprints the patterns of the mask in the resin, representing the outline of the electronic circuit, with its transistors, all the other components and the metal interconnection tracks.

Unless the mask is in contact with the resin, the patterns it projects onto the resin are bigger than those on the mask. Yet the aim is to produce smaller transistors. So a series of lenses is used to focus the light, somewhat as a magnifying glass concentrates the sun's rays. These lenses reduce the size of the projected pattern. All that then needs to be done is to dissolve the areas of insolated resin to expose the silicon with the exact outlines of the components. The silicon is then engraved with acid, as in Joule's day, though the technique has become more precise. With his knife blade, Joule was able to draw a line every 50 micrometers, in other words, one line every 50,000

---

[7]A microprocessor is a collection of electronic circuits incorporated onto the surface of a semiconductor, comprising a memory, a processing unit, synchronization circuits and a set of interconnecting wires.

nanometers. By the end of the 1960s, photolithography could do five times better and draw transistors occupying an area 10,000 nanometers square. By 2006, the distance between the transistor's input and output was no more than 90 nanometers. More than 80 million transistors crammed on an area the size of a fingernail! In 2007, when this book was written, integrated circuits using transistors measuring 65 nanometers between input and output came onto the market, and transistors measuring 45 nanometers were set to arrive in 2008. These devices are one hundredth the size of a red blood cell!

A Needle Upright on a Football Pitch

Photolithography has played a key role in this headlong rush to miniaturization. Just as it takes a fine brush to paint small details, it takes light with a short wavelength to engrave minute circuits on the surface of a computer chip. The resolution, i.e. the minimum gap between two distinguishable points on a circuit, depends on the wavelength. The shorter the wavelength, the finer the pattern reproduction, and the greater the resolution. So the wavelength of the light used has moved from visible light (400–800 nanometers) to ultraviolet (350–450 nanometers), then deep ultraviolet (220–310 nanometers), to reach 193 nanometers today. For the next stage (45-nanometer transistors), one possibility is immersion photolithography, a process in which a liquid is placed above the silicon, so that the mask appears from above to be enlarged. The liquid acts as an additional lens and makes it possible to produce smaller transistors using the same wavelength.

However, for the stage beyond (32-nanometer transistors), the engineers do not know what approach to take. They could reduce the wavelength of the light again, with a photolithography process using extreme ultraviolet with wavelengths of 13.5 nanometers. That would entail sophisticated and extremely costly optics to focus light onto the surface of the semiconductor. Another promising avenue is x-rays, since they have a very short wavelength of around a nanometer. The downside is that it is so short that x-rays can easily pass through matter, making the masks transparent, and x-ray optics are also very hard to control.

Another possible alternative solution to light and x-rays could be electron beams, which also have a very short wavelength. Electron beam lithography has been known since 1960, when Gottfried Mollenstedt, of Tübingen University in Germany, used an electron beam similar to those employed in scanning electron microscopes to score lines in a resin. He succeeded in drawing his university's logo with lines 100 nanometers thick. The principle is the same as with optical lithography: a polymer is chemically transformed when exposed to a beam of electrons. This transformation can be revealed with a solvent which dissolves the exposed parts, leaving very fine grooves. Currently, electron beam lithography is used to make photolithographic masks and, in research labs, to make the world's smallest transistors, with a distance between input and output of only 20, 15 or even 9 nanometers. They occupy the equivalent space on a chip to a needle standing upright on a football pitch. They tend to protrude above the surface of the semiconductor, like miniature mushrooms.

The First Limits to Miniaturization

However, it is not enough to be able to make ever-smaller transistors. They also need to work reliably. As they increase in number, the probability that one of them will be defective also increases. Engineers have applied a wealth of expertise and devised endless technological tricks to overcome all the obstacles encountered in this relentless pursuit of miniaturization.

One problem is the metal interconnect tracks between the transistors on the surface of an integrated circuit, which could extend over as much as 6 kilometers within a single square centimeter. As miniaturization progressed, these aluminum tracks became so narrow that the aluminum atoms were often swept away by the flow of electrons from an electric current, in a phenomenon called electromigration. Because it was difficult to manufacture kilometers of ultrapure aluminum, there were inevitably multiple impurities and grains on the tracks, which formed small resistances that created an electrical voltage when an electric current was passed through the wire. The resulting electrical field could become so intense that it gradually displaced the aluminum atoms, causing the track to disappear. This problem was resolved in 2001, by replacing aluminum with copper, which is more resistant to electromigration and is also a better electrical conductor. Following this change in material (which took fifteen years of research), electrons in integrated circuits now flow much faster.

Another problem arose when transistors began to be manufactured on the 65-nanometer scale. At this scale, the layer of insulation deposited above the transistor, separating the control electrode from the semiconductor "channel" (the link between the transistor input and output), is 1.2 nanometers thick. This means that it contains only five or six layers of atoms. It becomes a less effective insulator and there is a risk that electrons may leak out of the control electrode into the channel. As this leakage current increases, the insulator's resistance diminishes, as does the electrical field between the control electrode and the channel. Because this is the field that controls the opening or closing of the channel, it becomes difficult to control the movement of electrons in the transistor.

Historically, this insulating layer was made of silica (silicon oxide), which is a very good insulator when sufficiently thick. In order to resolve the leakage current problem, the silica had to be replaced by materials with better insulating properties, oxides of "rare earth metals," such as hafnium. This substitution reduced the intensity of the leakage current by a factor of 10, but the slightest change in the choice of transistor materials generates a host of knock-on effects. For example, the new insulator was not compatible with the metal of the electrode, which had to be replaced with a new metal alloy.

Leakage current is a quantum phenomenon, arising from the quantum properties of electrons. Cutting in at the 65-nanometer scale, it is the first sign of the disruptive impact of quantum effects, which engineers had generally ignored when designing transistors. From this point on, the continued miniaturization of transistors would require an understanding of such effects. This downshift resulted in the manufacture of new devices, at the 10–100-nanometer scale, different from transistors, which would give rise to other quantum phenomena. Before we move to this scale, however, let us look at the spread of these manufacturing techniques from microelectronics to other fields.

Contagious Miniaturization

Miniaturization spread inexorably beyond the field of electronics, reaching and transforming other sectors, like mechanical engineering. The machine tools used to make components by turning, milling or drilling had reached the limits of precision. They could produce excellent mechanical parts with a machining precision of around 1 micrometer, but that looked like the smallest they could go. In the 1980s, however, the University of California's Richard S. Muller was working on the optimization of manufacturing processes with silicon oxide, which was used as an insulator in integrated circuits. From his familiarity with photolithography processes, it occurred to him that these techniques could be used to manufacture a microwheel: he drew a wheel on the surface of a silicon wafer covered with silicon oxide, then engraved the wafer underneath in order to release it. This idea gave birth to the entire field of silicon micromechanics. Techniques drawn from microelectronics replaced the usual machining processes and cut the size of mechanical components from a few hundred micrometers to a few micrometers, with potential manufacturing precisions of a few nanometers. In turn, this silicon micromechanics gave rise to "microelectromechanical systems" (MEMSs), a combination of mechanical components (sensors or actuators) and electronics, in which detection or control signals are transmitted directly to the mechanical components. Like transistors in the microelectronics industry, MEMSs can be produced cheaply and in large quantities.

MEMSs are small-scale versions of the components and machines that exist on the macroscopic scale, fitted with wheels, pumps, valves, springs, clamps and gears a few micrometers — or even tens of a micrometer (100 nanometers) — in size. These mechanisms can be driven by micro-engines with gears the size of a red blood cell. MEMSs are found in inkjet printer injectors, in video projector micromirrors, and in the accelerometers used in video game motion detection systems. They are also present in photographic devices, video cameras, watches, pacemakers, and they account for 20%–40% of the cost of a car, forming components such as pressure sensors in the climate control circuit, braking force sensors, fuel level detectors, airbag sensors (which contain up to six accelerometers in top-of-the-range models).

MEMSs still have a long way to go. They are constantly building on advances in lithography techniques in the microelectronics field and have achieved great feats in research labs. For example, researchers have made silicon girders 100 nanometers thick — 1000th the thickness of a human hair — and 100 micrometers long. On our everyday scale, this would be the equivalent of a girder 1 meter long and just 1 millimeter thick, which is impossible.

Such girders (or microlevers) can, however, exist in the micrometric world. Instead of bending, they vibrate at very high frequencies. These vibrations are used to weigh molecules: when a molecule is dropped on the girder, its vibration frequency changes and the mass of the molecule can be deduced from this. The aim is not to measure the mass of the molecule, which is known, but to detect the presence of molecules. These vibrating girders can identify one molecule amongst a billion others.

Microelectronic manufacturing techniques and MEMSs have also transformed biology. Completely new systems for biochemical analysis have emerged, spearheaded by so-called DNA chips or biochips. These are made by grafting DNA strands onto silicon chips prepared using photolithography. Today, such chips can contain up to 300,000 strands of DNA, and are generally used to detect genome sequences for the purpose of identifying genetic diseases or viruses. The problem is that the DNA segments must go through a meticulous preparation process before they can be analyzed. It would be helpful to combine the different stages on a single chip, so that a raw sample (a drop of blood or water) could be analyzed directly. The objective is no longer to use silicon chips in laboratories, but to incorporate ultraminiaturized laboratories onto a chip.

Work is in progress to produce these so-called "labs-on-a-chip," which would be able to perform tasks such as providing the detailed composition of blood from a single drop. This will entail miniaturizing separators, chemical reaction chambers and sensors, and linking them by ducts finer than a human hair. An additional difficulty is to move the drop through these ducts. On this scale, surface tension phenomena dominate and the interactions between the drop and the walls prevent it flowing, as if it was "sticking" to the sides. Research is being done in microfluidics to develop microvalves or devices that apply intense electrical fields that can make a drop move through a microchannel.

Welcome to the Quantum World

Let us return to the world of integrated electronics and transistors. We have already looked at the problem of leakage currents in 65-nanometer transistors. In the ultimate transistors, on the 20-nanometer scale or below, currently being made in research labs, there is an even bigger problem. They do not really work. The malfunctions are not due to manufacturing defects, but arise from the architecture of the transistor itself. Its dimensions are so small that electrons inevitably escape from the control electrode into the transistor channel. To avoid such leakage, engineers are obliged to perform feats of acrobatics in the design and manufacture of the transistors. The production of transistors on the 45-nanometer scale is possible, but it will entail at least 400 technological steps. And after that? Like rats leaving a ship, these leaking electrons are an early warning sign: transistor miniaturization will soon hit the buffers. On this scale, the quantum behaviors that start to emerge in the electrical response or conducting properties of the minute electric wires are transforming microelectronics as it currently functions.

But what do we mean by the word "quantum," which has suddenly gatecrashed the technological party? In 1900, the German physicist Max Planck realized that exchanges of energy cannot take place continuously in the world at the bottom as they do in the macroscopic world. They must occur in leaps, called "quanta." These leaps are multiples of an elementary leap, the "quantum" (or quantum of action) which is a universal constant, called the Planck constant. For example, when an electron is part of an atom, its energy is defined in "quanta" because it can take only certain discrete values. The best explanation early 20th century scientists could find was to associate this electron with a wave. Indeed, the electron is trapped in its atom as if in a box. However, a wave trapped in a box reflects off the walls and therefore cannot possess all possible wavelengths, just as a guitar string stretched between the head and bridge of the guitar vibrates at a single frequency and emits a single note. Squeezing the string against the neck changes the note. Similarly, the energy of an electron cannot take on just any value in its little atom of a box, but is confined to certain discrete values. Another consequence is that we cannot precisely measure the position of the electron around its atom. We can only know the probability of its presence at a position in space.

Associating a wave with a particle like an electron has a major impact on the physics of electrons present in a solid. As soon as the size of one of the three dimensions of that solid (length, width or height) comes close to the wavelength of the wave associated with an electron, quantum behaviors occur. They were already present in larger transistors, but were masked, because the number of atoms involved in the composition of a "big" transistor is so gigantic that it jumbles the quantum waves, in much the same way as the instruments in a large orchestra would produce an undifferentiated noise if they all started playing different notes.

With smaller devices, the situation is different. The quantum phenomena are no longer masked. The study of such devices has opened up a new field of research: mesoscopic physics. They have dimensions ranging from a few tens of nanometers to a few hundred nanometers, at the intermediate scale between the atomic and the macroscopic, containing millions of atoms. In mesoscopic physics, therefore, the quantum waves associated with the electrons are still jumbled, but one of the jumbling factors has disappeared in comparison with larger devices. When the device becomes smaller than the average free travel of the electrons (the average distance they travel between two collisions), the probability that the vibrations of the atoms will collide becomes smaller, as if the electrons no longer had time to interact with each other. The vibrations almost cease and it looks as though a large number of electrons are associated with a single wave, as if a note had finally emerged from the hubbub of the orchestra as the musicians tuned their instruments.

Carbon nanotubes — which we will consider in the next chapter — are materials that lie at the frontier between the mesoscopic and the nanometric scale. Discovered in 1991, they are carbon tubes with a diameter ranging from a few nanometers to a few tens of nanometers. They can sometimes be up to a few micrometers in length, forming rolled sheets of graphite, like rolls of wire mesh. They have aroused great excitement, because they are very strong, can act as conductors or semiconductors, and have high heat conductivity. Researchers are looking into their potential use as electrical conductors or as channels for new types of transistors. They contain large numbers of electrons, and they are long relative to their diameter. When an electrical current is passed through them, it behaves classically, obeying Ohm's law. That is not the case in the cross-section of the tube where the order of magnitude is a few quantum wavelengths of the electrons. This means that the electronic properties of a carbon nanotube require an understanding of both the classical properties and the quantum behavior of these conduction electrons.

In mechanics, protein motors — assemblies of proteins in cells which transform chemical energy into work — also stand at this frontier between mesoscopic physics and nanophysics. Every protein in the protein motor is composed of thousands of atoms. The position of these atoms in space is governed by the laws of quantum physics. A quantum wave is associated with the way each chemical bond between two atoms in the protein vibrates. Since there are a large number of chemical bonds, the quantum wave associated with each type of vibration will not stand out from the global motion of the protein. Just as with electrons in a solid, the quantum vibration waves will be jumbled. The mechanical properties of protein deformation will then appear classical, in that the protein will be able to rotate or translate. The assembly of proteins forming the protein motor will thus seem to rotate classically, a phenomenon which is already being observed experimentally with *in vitro* preparations of these motors.

Mesoscopic physics, therefore, should not be confused with nanophysics. Nanophysics is about the physics of systems containing only a few tens of atoms and in which the jumble of quantum waves is either absent or controlled (or induced by the external environment). This will be the subject of the next chapter. However, as tends to happen with frontiers, there are conflicts between the scientific communities (see Appendix II). For proponents of the top-down approach who use microelectronic miniaturization techniques to do mesoscopic physics, nanophysics begins when quantum properties emerge. For those who practice a bottom-up approach and work with individual atoms, nanophysics begins when the atoms in a system can be counted one by one, and ends when there are too many of them and they create the first uncontrollable internal jumbles associated with quantum phenomena.

Pardon Me, Did You Say "Mesoscopic"?

In an ultimate transistor, the leakage current is caused by a quantum phenomenon called the "tunneling effect" (the effect exploited in the scanning tunneling microscope): electrons are present in very large numbers but are described by a single quantum wave. In the quantum world, the wave associated with a particle cannot suddenly stop at the surface separating two materials, so the waves maintain a degree of continuity. For this reason, there is a nonzero probability that the electrons will tunnel through from one material to another. Tunneling is a "walk through walls" effect, operating at a range of less than a few nanometers. In an ultimate transistor, when the insulating layer separating the control electrode from the active part of the transistor (channel) is less than 1 nanometer thick, there is a very high probability that certain electrons will tunnel through the insulating layer of the transistor from one electrode to the other or from one electrode to the channel. This electron "stampede" is a big problem. With each generation of smaller transistors, a new material will have to be found to provide an insulating layer that is more resistant to quantum electron wave penetration. In addition, the more extensively the control electrode covers the area of the channel, the better a transistor works. However, the intensity of the leakage current is proportional to the surface area concerned. Engineers are faced with opposing demands: to increase the efficiency of the transistor, they need to increase the area covered, but to diminish the leakage current, they need to reduce it. It is a catch-22.

Other quantum phenomena have been observed in minute devices, again produced using techniques drawn from microelectronics. On our scale, when an electrical current flows through a wire, it obeys Ohm's law: its intensity is inversely proportional to the electrical resistance of the wire; in other words, electrons flow less easily in a wire with high resistance. The narrower the wire, the higher the resistance. Electrons thus flow less easily in a narrow wire, like moviegoers caught in a bottleneck on leaving the theater.

However, once the diameter of a metal wire falls to a few tens of nanometers, a strange effect occurs: the resistance ceases to increase in inverse proportion to the cross-section. It increases in "jumps," remaining constant as the diameter of the wire shrinks, then suddenly rising, and so on, each step corresponding to an elementary "quantum" of resistance. To use the movie theater analogy, it is as if the moviegoers, instead of slowing gradually as they moved into the exit corridor, suddenly jumped onto a cart traveling at a constant speed for a few meters, and then slowed down in a narrower part of the exit corridor.

The reason for this step effect is diffraction. When the cross-section of the wire diminishes, at a certain point it becomes proportional to a few quantum wavelengths of the electrons. In this case, each time the cross-section of the wire becomes a multiple of a half-wavelength, resonance occurs and a quantum of resistance is attained. When the cross-section of the wire becomes less than the last half-wavelength, the wire becomes opaque to electrons. It is as if the diameter were too narrow for the quantum wave to move through it (just as light cannot travel through too narrow a hole), so the electrons can no longer flow. This quantum phenomenon was observed for the first time in the late 1980s.

What happens when a whole electrical circuit is made with very fine metal wires? In a classical electrical circuit, such as two resistors mounted in parallel, electrical conductance, which measures the "ease" with which current flows through a circuit, is the sum of the conductances of each component, according to Kirchhoff's laws (established in the 19th century). However, when this circuit becomes mesoscopic, i.e. when the size of the resistors and the diameter of the connecting wires fall below 100 nanometers, the circuit no longer obeys Kirchhoff's classical laws. Its conductance is the sum of the conductances of each resistor, plus an adjustment factor that reflects the effects of quantum interferences between the two parallel resistances. The basic electrical laws do not cease to apply, since the electron's charge is conserved. However, quantum effects arise.

The electron is a fundamental particle which has an electrical charge that is called elementary because it is, in principle, indivisible. By convention, this charge is -1. The electrical charges of all the particles of which matter is made are integers — multiples of the charge of the electron. When an atom gains an electron, its electrical charge is negative, because it contains one more electron. For example, the chloride ion of table salt (sodium chloride) is a chlorine atom that has gained an electron. Conversely, an atom that loses one or more electrons has a positive electrical charge, because it contains fewer electrons. The sodium ion of the same table salt, abbreviated to Na^+, is an atom of sodium (Na) with an electron missing.

If a conducting bar with a low current flowing through it is subjected to a magnetic field applied perpendicular to the direction of the current, a voltage occurs at the terminals of the bar. This voltage increases as the magnetic field increases. This is called the Hall effect. Hall resistance, which is the ratio between this voltage and the intensity of the current, increases linearly with the magnetic field. In 1998, physicists replaced this bar with a wafer a few nanometers thick. They observed that the increase in Hall resistance in relation to the magnetic field is no longer linear, but stepped. This is called the quantum Hall effect.

The first step represents elementary quantum Hall resistance; the subsequent steps are multiples of this elementary resistance. With an even higher magnetic field, other intermediate steps appear. Where do they come from? This was a real headache for physicists. In fact, these intermediate steps are explained by the existence of "carriers" with a charge less than — in fact a fraction of — the charge of the electron. Nevertheless, they are not caused by "fragments of electrons." The experiment does not undermine the structure of the electron or its status as an indivisible particle. These carriers of a divided charge are particles of a new kind, called quasi- or virtual-particles. They are the result of the collective behavior of thousands of electrons present in the ultrathin wafer. In this wafer, under the influence of the magnetic field, it is as if the current were carried by quasi-particles with a one-third charge. Other experiments have confirmed the existence of these quasi-particles, and have revealed others with a charge that is 1/5 or 1/7 of the charge of the electron. In this sense, miniaturization is taking us into a new world.

The Electronics of Tomorrow

Skeptics had already predicted insurmountable barriers to the production of transistors for technology at the 1-micrometer scale, then a quarter of a micrometer, and finally 100 nanometers. Scientists found ever-more-ingenious solutions to skip cheerfully across all these frontiers. However, for the generations of sub-20-nanometer transistors, quantum properties are technically unavoidable and the very concept of the transistor comes into question. Further miniaturization is still possible, but these devices will have little in common with existing transistors. What will they be like? Different avenues are being explored, which make use of, rather than overcome, quantum effects. A new kind of electronics — called quantum electronics — is beginning to emerge in research labs. This field is sometimes — wrongly — called nanoelectronics. The fact is that such devices still contain thousands of atoms. It is only the manufacturing precision that lies at the nanometric scale. Let us take a look at some of the future directions for electronics.

At around the 20-nanometer mark, classical transistors have fewer and fewer "active" electrons (electrons that affect the operation of the transistor). So why not make a switch that works with a single electron? When an electron enters a standard transistor, the internal energy of the transistor increases. It draws this energy from the thermal fluctuations of its point of "origin," and because the increase is minute, it is lost in the thermal fluctuations of its "destination." One electron more or less in the transistor makes no difference. However, when the transistor is no more than a dozen nanometers wide, this additional energy ceases to be negligible, since the number of thermal fluctuations is small. In addition, if an electron manages to enter the transistor, its presence is enough to prevent others entering. This phenomenon is called Coulomb blocking: no more electrons can enter, because the energy cost is too high. It is like a lock on a canal — once the lock is full, no more water can enter. Coulomb blocking was described in 1951, but the idea of using it to make a single-electron transistor emerged in 1985, and was then applied two years later using cutting edge electron beam lithography techniques.

Another possibility is an entirely different kind of electronics based not on the charge of the electron, but on its intrinsic "spin." Like a small magnet, this spin produces a magnetic field that points upward or downward, depending on the direction of rotation. In a nonmagnetic material, the direction of spin is random. So, in a classical transistor, the direction of spin of each electron is random and does not affect the transistor's properties. In a magnetic material, however, the number of electrons that point upward and downward is different, which gives the material its magnetization. So, when electrons travel through this magnetic material, their spins interact with the material's magnetic moment. For example, when any electrical current (in which the directions of spin are random) passes through a ferromagnetic material, electrons with a particular direction of spin are more likely to get through than others. The ferromagnetic material acts like a filter and a large proportion of the outgoing electrons have the same direction of spin.

Another possibility: electrical current can modify the direction of local magnetic moments in a magnetic material, or else the application of a voltage can change the direction of the material's magnetic moment. This material constitutes a "spintronic" component. The change of spin is quicker and requires less energy than shifting an electronic transistor from one state to another (open or closed), because the spins flip more quickly than electrons flow. That is why many scientists have come down in favor of spintronic electronics.

Of all the candidates that might replace the miniaturized semiconductor transistor in coming years, the single-electron transistor and the spintronic transistor have the advantage of using manufacturing technologies that have already been developed for the miniaturization of classical transistors. Whatever solution emerges, any of these devices will exploit quantum effects rather than trying to get round them.

The Guiding Thread

The myth of nanotechnology is that Feynman's supposedly visionary foresight led to ultraminiaturized transistors, DNA chips and micromechanics. In fact, these nanotechnologies are the logical outcome of standard techniques developed in the late 1950s, such as photolithography for microelectronic and micromechanical components or electron beam lithography for mesoscopic physics. Nanotechnologies sometimes produce objects a hundred times the size of a nanometer, with a manufacturing precision of a few nanometers. Yet they occupy center stage, ineluctably associated with the extremely small, whilst another technology is developing at the scale below, based on the manipulation of single atoms and generating devices with dimensions of a few nanometers and manufacturing precisions of 0.1 nanometers. This technology will be the subject of the next chapter.

Apart from the technical feat of fitting ever-larger numbers of transistors onto the surface of a semiconductor, and its economic and practical contribution to our everyday lives, there is another guiding thread to the adventure of miniaturization, this time in the realm of metaphysics. Will we one day be able to make a machine that thinks? When Pascal incorporated a calculating machine into nonliving material, his "pascaline" did not think. When Watt designed his miniature steam engine, he added a program in the form of a series of holes drilled in a small metal plate, which controlled the machine's valve sequencing. Yet a steam engine does not think. When, in 1820, Babbage designed the first universal mechanical calculating machine, it was still not a thinking machine. Today, when engineers manufacture a chip with 100 million transistors, it is clear that the chip does not think. How many cogs, vacuum tubes or transistors does it take for a machine to start thinking? In 1957, John von Neumann suggested 100,000 transistors. With miniaturization, that figure was exceeded a long time ago, but ... machines still do not think.

CHAPTER 3

STAYING AT THE BOTTOM

One day, miniaturization will hit the buffers. The time will come when the tiny fragment of material will be too small to accommodate a device or even a machine. In the 1960s, the ultimate limit of miniaturization was thought to be around the scale of the macromolecules of life, like DNA proteins, which are made up of thousands of atoms. This was the time when the powers of these macromolecules — their ability to store information, transport molecules, produce energy and communicate — were being discovered. There are even some enzymes[8] that have several active chemical sites and are regulated by other molecules, which means that they are active only in the presence of a molecular or electric signal, somewhat like electronic relays. In 1970, in *Chance and Necessity*, the 1965 Nobel Prize-winner Jacques Monod wrote, as if throwing down the gauntlet to physicists, that the minimum weight of an electronic relay was around 10^{-2} grams, whereas the weight of an enzyme capable of the same performance was some 10^{-17} grams, a million billion times less. The point he was making was that the living world could outperform our ultraminiaturized devices. At that time, it was inconceivable that we might make machines smaller than macromolecules. Today, however, we know that these biological entities are not the smallest objects capable of embodying a device. The challenge laid down by Monod has bounced beyond the techniques of miniaturization — a single molecule is a material entity that has sufficient spatial extension and temporal stability to embody a device or a machine. Hence the new idea, which took shape in the 1990s, of reversing the way machines are made: instead of starting with a mass of material and whittling it down to the smallest possible dimensions, why not start with a few atoms and combine them to make a machine with

---

[8]Enzymes are macromolecules which have the role of accelerating a chemical reaction millions of times.

just the right number of atoms? This idea is the basis of a new technology —
nanotechnology. The subject of this chapter is the first stage in this bottom-
up approach to complex molecule machines: we will begin by "staying at
the bottom" of the scale of magnitudes and discovering how it is possible to
manipulate an atom or a molecule consisting of a few dozen atoms, entities
much smaller than biological objects.

An instrument invented in 1981 — the scanning tunneling microscope
(STM) — opened the door to the world below, by providing a technical
solution to this technological U-turn. The first picture of a single molecule
was produced in 1957 with an electron microscope (see Appendix I).
However, it was the STM that first made it possible not only to display the
picture of a molecule on a screen, but also to touch that molecule with the
tip of the microscope. The instant the molecule acquired the status of a gen-
uinely independent material entity, the adventure of nanotechnology began.
With it, devices have been manufactured with dimensions smaller than anything previously achieved, in the region of 1 nanometer, with manufacturing
precisions of some one-tenth of a nanometer. Nanotechnology is a new stage
in the long saga — not of materials science, but of the science of matter.

Birth of the Molecule

The tip of the STM can turn a molecule into the tiniest machine in existence. Originally, however, the concept of the molecule was proposed only
as a solution to the problem of the identity of bodies. The molecule was
defined as the smallest part characterizing a substance. It is a subject that
has been hotly disputed. Giovanni Alfonso Borelli (1608–1679) thought
that a substance was an aggregate of "little machines" (*machinulae*) capable
of combining or separating. Seventeenth century scientists recognized the
inadequacies of the then-dominant Aristotelian doctrine, according to which
matter was continuous and made up of four elements — earth, air, fire and
water. One of these thinkers, the Dutch physician and mathematician Isaac
Beeckman, who maintained an assiduous correspondence with the scientists
of his time, recorded his thoughts and experiments in a scientific journal.
On September 14, 1620, he wrote that if a medicine is divided in half, each
of the parts retains its curative powers. If it is divided again, the same is true.
However, he reasoned that with continued division, a time comes when the

infinitesimal fragment loses its properties. Beeckman called the smallest part that retains its therapeutic efficacy the "minimum" — the equivalent of the modern molecule. He imagined that it is composed of atoms, all made up of the same "raw material," but differing in their "shape." He explained that there are at least four types of atoms (corresponding to the four elements), but perhaps more (today we know of 118).

In 1621, Sébastien Basson, a theologian and teacher at the Die Academy in Dauphiné in France, formulated a similar concept. He was examining the origin and constitution of matter by studying pre-Aristotelian sources which, being closer to the Creation, were therefore necessarily more reliable. He examined the arguments of the atomists, for whom matter is discontinuous and composed of atoms, including the argument based on the experiment of pouring a drop of wine into a large volume of water. The wine disperses and gradually disappears, which, according to the atomists, shows that matter is divided into particles. Basson asserted that matter consists of primary particles, which he also called minima. His minima are composed of the four elements and differ from each other in the proportion and arrangement of those elements. He imagined that the minima assemble to form secondary particles, which collect into tertiary particles, and so on, eventually forming the objects that surround us. This was the birth of the concept of the molecule, the smallest characteristic part of a body, but a part composed of other things (elements). However, the word "molecule" (*molecula*) appeared later, in 1636; it was coined by the French priest Pierre Gassendi, from the suffix "-cule" and "mole" (the equivalent of today's "mass"), in translation of the word "particle" from a piece by Diogenes Laertius on the atomist philosopher Epicurus (the word *molecula* did not correspond to the molecule as we understand it today).

Although hypothetical, molecules were to transform the science of matter. Antoine de Lavoisier (1743–1794) showed that a body retains its identity, whether in gaseous, liquid or solid form. Steam, water and ice are the same body, simply with a different arrangement of molecules in their different states. Lavoisier was a craftsman of the "molecularization of the image of the world"[9]: the concept of the molecule developed to such a

---

[9]Henk Kubbinga, *L'Histoire du concept de molécule* (Springer, Paris, 2002).

degree that by the end of the 18th century scientists sought to explain all phenomena in terms of molecules.

In the 19th century, the science of matter progressed as never before. The Englishman John Dalton argued that matter is composed of atoms of different masses, combined into molecules — the first time that this accurate description of matter had been put forward. The Italian chemist Amedeo Avogadro then showed that two bottles of the same size, with identical temperature and pressure, contain exactly the same number of molecules (approximately 27 thousand billion billion molecules per liter), whatever the gas in the bottle. The molecule was truly taking shape. However, each scientist had his own language. Avogadro spoke not of atoms but of "elementary molecules," whereas Dalton called molecules "compound atoms." A big conference was organized in Karlsruhe in 1860 to clarify the situation. Following heated discussions, the chemists came to an agreement on fundamental definitions, which are those still in use today. The agreement on the distinction between the atom and the molecule (a group of atoms) was ratified.

So How Big is a Molecule?

From this point, the race was on to determine the size of a molecule, bearing in mind that its existence was still only a hypothesis. The Austrian physicist Josef Loschmidt (1821–1895) calculated the diameter of a "molecule of air" and came up with 9.69×10^{-7} millimeters, i.e. 0.969 of our current nanometers, which would have been a very good result... if there were such a thing as a molecule of air.[10] Using another method, the English physicist Lord Kelvin (1824–1907) estimated the size of zinc and copper atoms at 0.1 nanometers, roughly the right order of magnitude. Long before this, Benjamin Franklin (1706–1790) had inspired an experiment which, a century later, was used to calculate the size of a molecule. Along with others, he had noticed that oil spreads across the surface of water and forms a film. Assuming that the film is one molecule thick, you only have to divide the volume of oil by the area of spread to find the size of the molecule, which is around 1 nanometer (this is still a common classroom experiment today).

---

[10] Air is made up of molecules of different gases: nitrogen, oxygen, hydrogen, etc.

Still, in the 19th century, chemists were confronted with a mystery: substances could be composed of exactly the same molecules yet have different properties. How was that possible? The Swedish chemist Jöns Jacob Berzelius came up with a suggestion: "In the future, perhaps, this [enigma] will be resolved by studying the geometrical shape of [molecules]." In the meantime, he gave the name "isomers" to these embarrassing chemical species. He was right: in 1875, the chemists Jacobus Van't Hoff and Joseph Le Bel showed that the bonds on a carbon atom point from the center of the atom to the corners of a tetrahedron. The molecule extends in the three spatial dimensions. This shows that two molecules made up of the same atoms can adopt different architectures and therefore have different properties. According to the German physicist Rudolf Clausius, this architecture is not a rigid whole — its atoms can oscillate slightly in relation to each other, even in a solid. In 1890, the young German chemist Hermann Sachse went further and demonstrated that molecular architecture is not always fixed, but adopts different conformations, as if it were "pliable." Finally, at the end of the 19th century, the molecule had come to resemble what we see today: a lattice of atoms able to adopt different configurations in space.

Scientists were now able to explain many macroscopic phenomena with molecules, but nobody had ever seen one, since they were much too small to show up under an optical microscope. They therefore remained hypothetical and, even at the end of the 19th century, there were still eminent scientists who continued to reject the concept of the molecule. In France, for example, the ubiquitous Marcelin Berthelot, a man of great scientific and political clout (Professor at the Collège de France, Member of the French Academy of Sciences, Minister of Public Education and Minister of Foreign Affairs), could refer to the molecule as a "mystical concept." The existence of molecules was finally established in 1908, thanks to the French physicist Jean Perrin, who provided experimental proof.

Maxwell's Demon

In 1871, the Scottish physicist James Clerk Maxwell triggered a genuine, but initially unnoticed, revolution. In an unusual thought experiment, he imagined a being sufficiently small to be able to measure the speed of every molecule of a gas enclosed in a box. This "little demon," as it would subsequently be called, would be capable not only of "following" the molecules as they moved around at high speed and in all directions, but also of dividing them according to speed, slow on one side, fast on the other. So, if the demon was allowed to work in a room at normal temperature, half the room would fill with ice cubes whilst the other half would become scalding hot, a reflection of the fact that temperature is directly linked with how fast gas molecules move. Maxwell was conducting a thought experiment, but it was also the first instance of a complex device operating at the molecular scale. In the 1870s, the molecule had only just acquired a shape, so it was inconceivable to design such a molecular device, let alone to make it. Maxwell's demon continued to intrigue thermodynamic theorists, but the underlying concept of a molecular machine was shelved for almost a century.

In the 20th century, biologists, notably the Frenchman Jacques Monod, brought Maxwell's demon out of the closet. Their goal was to explain the elementary processes of life through macromolecules, some with the role of maintaining the solidity of the cellular structure, others playing a more active part. In 1947, for example, the Hungarian biochemist Albert Szent-Györgyi, the discoverer of vitamin C, suggested that proteins guide electrons around their atomic skeleton, a little like an electric wire. The version of the molecule depicted in this account is not yet as complex as Maxwell's demon, but in assigning it the function of a guide wire, Szent-Györgyi recognized that an electric wire does not have to be made of an extruded metal. A single molecule can perform a particular role in a complex assembly. A single molecule could be used, as Maxwell had suggested, to do physics. Thus rejuvenated, the demon continued along its scientific path into the hands of the chemists.

In the late 1950s, Henry Taube, an American chemist, took up Szent-Györgyi's idea in his experiments, designing and synthesizing elongated molecules (1 nanometer long and 0.2 nanometers in diameter). He even performed a spectroscopic analysis, which is to say that he measured their light absorption by passing a ray of light through a solution containing billions of these molecules. He managed to detect infrared absorption that could only be generated by an intramolecular phenomenon, and showed that this energy is used to transfer an electron from one end of the molecule to the other. This demonstrated that a molecule can steer electrons one by one, almost like an electric wire. For the first time, a molecule had been designed to direct electrons from one end of its structure to the other.

In the early 1970s, the American chemist Ari Aviram was working at IBM's T. J. Watson laboratories near New York, completing his doctoral thesis with New York University's Mark Ratner. The focus of their work was an idea which physicists of the time found hard to accept, and which would intrigue electronics specialists for a long time. They designed a device more functional than an electric wire — a one-molecule electric rectifier, i.e. a molecule that would let electric current flow in only one direction. To do this, they devised a molecule 1.2 nanometers long, made up of one electron-rich section and one electron-poor section. Once connected to an electrode at each end, the molecule would function like an electric rectifier: electrons would be unable to penetrate the electron-rich side (they are repelled), but could easily enter the electron-poor side. They would flow through the body of the molecule, allowing current to flow in only one direction.

A hundred years after Maxwell's demon, therefore, Aviram and Ratner had described a molecule able to act autonomously, an ultraminiaturized device in its own right. In the way it works, it is similar to Maxwell's demon, since it sorts not molecules but electrons, allowing them to flow in only one direction. But the main thing is that this molecule does not have to be stacked with billions and billions of others like it to perform a function. This idea of turning a single molecule into an electronic device marks the beginnings of molecular electronics. However, one problem remained: How do you connect macroscopic electric wires to a molecule?

How to Connect a Molecule

In the mid-1980s, there was no solution in sight for running an electric current through a single molecule. Aviram and Ratner were finding it hard to confirm whether their molecule really did rectify current. There was even a widespread belief that a molecule was much too small to be connected to an electric wire. Others simply could not see the point of molecular electronics, so successful was the miniaturization of microelectronic components. The idea of molecular electronics was running out of steam.

I can still see myself in 1985, as a young researcher with a brand-new doctorate, putting the solution I had in mind for resolving this problem of connection to the director of the Electronic Optics Laboratory at the National Scientific Research Centre in Toulouse. My idea was simple: if we pointed the beam of a highly focused electron microscope at a molecule, we might be able to insert a few electrons into it, then collect them in an electrode connected to the other end of the molecule. The idea was too simplistic and the director convinced me that the molecule would be fried by the electron beam before it could catch any electrons.

On his side, Aviram had also discussed this question at length with IBM's specialists in electron beam lithography. His plan was as follows: since it was already possible at that time to use electron beam lithography to produce a 20-nanometer-wide metal electric wire on a silicon surface, why could not they slice this wire in the middle and slip a few molecules into the resulting gap? Unfortunately, the specialists thought that it would be impossible to do this with the necessary precision.

In the meantime, the technology of the brand-new STM, invented at IBM's European research laboratory in Zürich, was spreading through all of IBM's research labs and a few university laboratories. In 1983, STMs were installed in the lab where Aviram worked. Physicists used it to observe the surface of semiconductors at atomic resolution, in order to understand their structure and the structure of their defects. Aviram was not yet involved, as he was busy synthesizing a new molecule.

He was developing a molecule which could change shape to act as an electric switch. The principle was as follows: the molecule is positioned at the junction of two metal electrodes and an electric field is applied which transfers two hydrogen atoms along the molecule, altering the molecule's electronic structure, i.e. the distribution of the electrons inside it. This modifies its electric conductance. Since the conductance is greater in the new configuration, the current in the external electric circuit should increase. Because the transfer of the hydrogen atoms is reversible, the molecule acts as an electric switch. However, as this device was even smaller than the rectifier molecule, Aviram had a hard time convincing his IBM colleagues of the practical viability of his work.

In 1986, he became aware of the potential of the STM and realized that its very sharp tip (only a few atoms wide) might act as an ultraminiaturized electrode that could be connected to a single molecule. It was at this time that I joined his research team, working on establishing a connection between the tip and a molecule. We began by fabricating gold surfaces, on which we scattered the switch molecules that Aviram had just synthesized. These surfaces had to be ultraflat, so that ordinary asperities would not be mistaken for our molecules. However, when we moved the tip toward the surface, we encountered a problem: the tip drifted and usually remained directly above the molecule for much less than a second, which was not long enough. If electrons were being transferred from the tip to the molecule, the STM's control electronics did not have time to record the signal. So we modified the control electronics to speed up the rate at which the electric signal produced by the molecule under the tip was recorded. And crossed our fingers.

This time, the tip remained steady above a molecule long enough for us to record its electric characteristics! Unfortunately, they remained standard and showed no signs of the switch we were hoping for. However many times we ran the tip over the surface and identified the position of the molecules, we failed to get the jump in current characteristic of a switch. Finally, after even more careful preparation of the tip, the increase in the electric field between the tip and the surface triggered a sudden change in the intensity of the current. We had succeeded in running an electric current through our molecular switch!

But our euphoria was short-lived. We realized that, during the experiment, a few gold atoms had in fact broken free of the surface and had short-circuited the junction between the tip of the microscope and the surface. What we had taken for a current caused by the closing of our switch molecule was in fact an atomic short circuit. It would take a further ten years of improvements to the STM before an electric connection could be established between its tip and a molecule. At present, that is the only reliable way to connect to a single molecule. Nonetheless, this experiment had shown that an STM could be used to connect to a few atoms. The experiment brought molecular electronics into the era of nanotechnology and prompted a resurgence of interest in this field of research.

Man Moves Atom

Many scientists had long been skeptical about the possibility of connecting to a molecule — i.e. exchanging electrons with it — because of the quantum properties of electrons. It was thought that electrons near and in a molecule would behave with essentially quantum randomness. So how could they be controlled? In establishing these principles, the inventors of quantum mechanics had played a cruel trick on experimenters. Moreover, according to Schrödinger, because of the quantum nature of the physical properties of the atom, there was no way to pinpoint the position of the wave associated with an atom, and therefore to manipulate an atom like a solid object.

The first images of a tungsten atom, obtained in the 1950s by Erwin Müller using a field ion microscope (see Appendix I), had generated a wave of scientific controversy. Some came close to concluding that the quantum theory of matter was false! Others, conversely, wondered whether the pictures obtained by Müller were not generated by parasitic phenomena or interferences.

In the early 1970s, students of Müller managed to make an atom jump from site to site on the tip of a field ion microscope, using the electric field and the temperature of the tip. They were able to follow the trajectory of this atom almost live on screen, as it moved randomly on the surface of the tungsten tip. This experiment should have ended the controversy. Despite Schrödinger's claims, it showed that it was possible to pinpoint the position of an atom and even to track its movements! But the controversies continued.

They only came to an end in the winter of 1989, thanks to the relentless efforts of Donald Eigler, a researcher at IBM's Almaden laboratories in California. Eigler had just spent two years at the Bell labs on the east coast of America, where the emblematic component of electronics — the transistor — had been born. There, he had begun building an STM to observe how a rare gas like xenon interacts with a metal surface. It was a question that plagued him. While working on his doctorate at the University of California at San Diego, he had researched rare gases, employing the standard technique of projecting them onto the surface of a metal to study their magnetism, which provides information on the metal's surface electrons. At Almaden, Eigler continued building his ultrastable, very-low-temperature STM, spending three years on the project. Instead of projecting xenon atoms, he deposited them on the surface and observed them, still with the aim of studying their interaction with a metal surface. Because the atoms of rare gases like xenon are chemically very stable, they showed very little interaction with the surface and escaped. Eigler had to cool the surface to a very low temperature to stop the atoms getting away.

One night (vibration levels in the building being lower at night), Don Eigler was shooting a sequence of pictures of the same part of the metal surface. The tip of the microscope would continually scan the surface, and each scan was recorded on a video recorder. The next day, as he fast-forwarded through the recorded images, he noticed that the atoms had moved from one image to the next, in the direction of the scan. He repeated the experiment and showed that, depending on the voltage and current applied to the tip, he would obtain either an ordinary image or a modified image in which the atoms had moved. He showed that the atoms had not moved randomly overnight, but could be deliberately and reproducibly manipulated, against all expectations and counter to the precepts of quantum theory. To confirm his discovery, Eigler wrote the abbreviation "IBM" using 35 xenon atoms. The picture flashed round the world and marked the birth of nano-technology: man had just "walked on the atom."

What was going on under the tip? The xenon atom can be compared with a soccer ball on a grass soccer pitch. The ball stays still, because it is wedged by the blades of grass. When a player places his foot on top of the ball and applies a little pressure, he traps it under his boot. If he moves his foot while maintaining the pressure, the ball rolls and follows his movements. However, if too much pressure is applied, the ball slides away. It is the same with a xenon atom under the tip of an STM. To get a good picture of a xenon atom without moving it, the tip must be held more than 0.2 nanometers away (the equivalent of a soccer player holding his foot just above the ball). But when the tip is held less than 0.2 nanometers above the atom, an interaction is created which alters the connection with the surface. The atom is "trapped," and this trap moves when the tip of the STM moves.

Eigler went on to draw different patterns with metal atoms and even with small molecules, such as a 5-nanometer-long "molecular man" produced with carbon monoxide molecules. His technique generated great interest in Japan. Hitachi's management asked its researchers to find a way of writing with atoms. Instead of drawing letters by depositing atoms one by one on the surface of a metal, they developed a technique for removing atoms one by one from the surface of a semiconductor: the interstices formed letters. Hitachi responded to Eigler's "IBM" with "PEACE'91 HCRL" (Hitachi Central Research Laboratory).

The technique of manipulating atoms discovered by Don Eigler was quite simply hijacked politically to justify a big drive to regenerate research in the USA, then in Japan and the rest of the world, as we explained in Chapter 1. However, until the mid-1990s, no other laboratory repeated Eigler's experiment, because no other lab in the world had an STM as stable as his. Then, following the advances achieved by Gerhard Meyer at Berlin Free University, STMs capable of manipulating single atoms came on the market (at a cost of around €0.5 million).

And Yet It Moves!

Don Eigler's pioneering work raised new questions. For example, is it possible to manipulate large molecules one by one? As with atoms, the tip can "press" on the molecule, trapping it. In the case of molecules, however, the trapping energy is dissipated by the chemical bonds between the atoms making up the molecule. The result is that the molecule does not move or slides away from the tip when the experimenter presses too hard.

Together, Jim Gimzewski, a physicist at IBM's research lab near Zürich, and I found a solution to this problem. He was one of the young physicists being encouraged by IBM to use the STM in all fields of physics and the chemistry of surfaces. The instrument could be employed to observe phenomena occurring on the surfaces of metals and semiconductors, such as how a boron atom (utilized to dope semiconductors) enters the semiconductor lattice and distorts it. The microscope generated superb and highly informative images. IBM was keen not to be outpaced. While I was struggling to run an electric current through a molecule with Aviram in New York, in 1988 in Zürich Jim had obtained the first pictures of a large molecule, phthalocyanine (a pigment), deposited on a silver surface (see Appendix I).

He continued his work imaging large molecules and I helped him to understand how those images were formed. The STM (see Appendix I) forms an image by using the cloud of electrons surrounding atoms, rather than the atoms directly. The more transparent this electron cloud is to the tunneling electrons, the stronger the signal which the microscope provides. It establishes a map of this transparency, which is proportional to the electric conductance of the "tip–molecule–surface" tunnel junction. Converting this map into a synthetic image — the "photograph of the molecule" — is not an immediate process. In fact, it can often be difficult to deduce the morphology of the molecule from this map, or even simply to recognize it.

In 1995, we were studying a large molecule — porphyrine — for which we had obtained conductance maps, but were unable to understand precisely what they represented. Jim was recording the images with Thomas Jung, a young physicist then a member of his group, and I was doing the calculations to interpret them. In April, I received an e-mail from Thomas: "It moved!"

Not long before, Jim and I had decided to introduce an additional parameter into the experiment by raising the body of the molecule slightly above the surface in order to alter the interaction between them. We wondered how this would affect the conductance map and were running new experiments with a porphyrine molecule fitted with four little molecular legs that raised it 0.4 nanometers above the surface. Thomas was to record a series of pictures of these four-legged molecules. Like Don Eigler, not wanting to spend the whole night in front of his computer screen watching the pictures one by one, he decided to record them on video. Fast-forwarding through the images the next morning, he realized that some of the molecules on legs had moved in the direction of the scan. That was when he sent the e-mail. We had discovered how to move a large molecule with the tip of the microscope, by first fitting the molecule with legs. This might seem self-evident. In the early 1990s, however, no one knew whether or not it was possible to apply the same mechanical concepts to a subnanometer-sized object as to macroscopic objects. We were so impregnated with the principles of quantum mechanics that we did not dare to use classical mechanics at the nanometric scale, on a single molecule.

Yet the molecule had undoubtedly moved in accordance with the laws of classical mechanics. We showed, by numerical simulation, that — provided that the legs are long enough and that the tip is at the right height to interact primarily with the central board of the molecule — part of the pushing energy generated by the tip is not absorbed by the molecule, but propels it forward. This required the height of the tip to be very carefully controlled. Moreover, it was no longer necessary to cool the metal surface to move a molecule, as it was with xenon atoms. Indeed, the legs of the molecule interact sufficiently at four points on the surface for it to remain obediently in place, without escaping, even at room temperature.

Since then, a large number of molecules have been moved around on metal or semiconductor surfaces. The technique of molecular manipulation is now well understood. However, other questions arise. Is it possible to manipulate atoms and molecules on the surface of an insulating material? On the surface of a metal or a semiconductor, the tip, the molecule and the surface interact together. The surface therefore contributes to the trap where the tip holds the atom or molecule. On an insulating surface, these interactions do not happen and the trap no longer works. A number of teams are studying this question, working on weaker, so-called van der Waals, interactions. The next question is much more futuristic. For the moment, atomic and molecular manipulations are restricted to the two dimensions of a surface. Is it possible that one day we might be able to detach a molecule from the surface and move it at will in three dimensions?

In any case, until we are able to manipulate an atom at will in space, the tip of the STM provides a key to unlocking the laws that apply in the world at the bottom. Atomic manipulation allows us to conduct unprecedented experiments — for example in the physics of electricity or in mechanics — on a single molecule.

The First Experiments in Nanophysics

At the macroscopic scale, an electric switch generally consists of a moving metal component fitted with a return spring. When tilted, this component connects two electric wires. In the world at the bottom, the smallest possible tilting component is an atom. Back in 1987, Aviram had created a switch molecule and we had made the first attempt to establish an electric connection to that molecule. Instead of a connected molecule which has to change shape to allow a current to pass, in 1993 Eigler proposed a simpler solution: using a single xenon atom as the tilting component. He controlled its movement by applying a pulse of voltage between the tip and the surface: the atom transferred between the tip and the surface at will. In its absence, the intensity of the current was very low, which meant that the switch was open. When the atom tilted under the tip, the intensity of the electric current increased fiftyfold, showing that the switch was closed. Since this transfer was reversible, this device was the world's first atomic switch. Ten years later, Francesca Moresco of Berlin Free University made a switch using a molecule as the tilting component. Compared with an atom, the advantage of using a molecule is that the return force can be controlled: the chemical structure of the molecule governs its interactions with the surface, speeding up and slowing down the switching process.

The next experiment in nanophysics focused on establishing an electric connection to a molecule. In 1987, Aviram and I had defined an experimental procedure for connecting a molecular switch. In this procedure, molecules are dispersed on a metallic surface that acts as a first contact electrode. The experimenter moves the tip of the STM toward one of these molecules, so that the tip acts as the second electrode. The tip is then moved slowly toward the molecule to establish an electric contact. The question is: How do you know when the connection between the tip and the molecule is established?

As the tip descends, it deforms the molecule. Too far, and it crushes the molecule. In fact, the current passing through the molecule increases as the molecule changes shape: electric contact with a molecule is established at the maximum current at which the shape or the electronic structure of the molecule remains unaltered. The trick is to find a balance between maximum current and minimum deformation, which requires great control of the height of the tip and was only achieved in the mid-1990s, when Jim Gimzewski and I were making another attempt to connect the tip with a molecule, this time fullerene (a molecule shaped like a soccer ball with 60 carbon atoms). The experiment worked as follows. We deposited a few molecules of fullerene on the surface of a gold crystal. We gradually moved the tip above one of the molecules and recorded the current in the electric circuit formed with the gold surface, the fullerene molecule and the tip at different tip heights. After a steady variation in the intensity of this current, we observed a sudden increase when the tip was positioned 1.1 nanometers from the surface. By raising the tip slightly until it was at the threshold of this tipping point, we could be certain that the molecule was not deformed and that an electrical contact had been established with the fullerene molecule. For the first time, we had made an electrical connection with a single molecule.

Having established this connection, we measured the molecule's electrical resistance. This "electrical resistance" is defined only in relation to the electrodes, i.e. the surface and the tip, and is not intrinsic to the molecule. A year later, Eigler would use the same principle to measure the electrical resistance of the world's smallest electric wire, consisting of two xenon atoms. This was the start of electrical experiments using a few atoms or a single molecule.

The Mechanics of a Molecule

Let us move on to the first mechanical experiments using a single molecule. We have already described how to push a molecule with the tip of an STM. In 1998, a chance discovery was to launch the era of "nanomechanics." But let us take a step back.

In the late 1960s, an American biochemist called Paul Boyer suggested that a protein can change shape through the rotation of one of its parts, i.e. that in the world at the bottom, a macromolecule can rotate on its own. In that case, might it be possible to use it for mechanical purposes? In 1997, the Japanese scientist Kazuhiko Kinoshita and his team managed to observe this rotation on screen, after attaching a fluorescent marker to the spinning part of the protein. Boyer's suggestion and Kinoshita's observation applied to macromolecules composed of thousands of atoms, but what was to stop us observing the rotations of a single molecule?

At this time, Jim Gimzewski and I were studying the way that relatively flat molecules of decacyclene assemble on the surface of a copper crystal. The decacyclene molecule is made up of a central benzene (flat hexagon) with six little "legs" attached to it. Our initial idea was to continue exploring ways of imaging a molecule at different distances of the molecule body from the surface. The legs of the decacyclene molecule were shorter than the legs of our first porphyrin molecule. Our experiment consisted in evaporating enough molecules on the surface to cover it with a single compact layer, in which all the molecules had their place in a very regular structure. However, the arrangement of the molecules across the surface was not always perfect. There were defects in certain places, such as a molecule missing or out of place. These defects created small spaces, some of them roughly the size of a molecule. What happened to a molecule in the layer near one of these empty spaces? On occasion, it moved as if trying to explore the space.

Luck was on our side, because in some of these spaces the molecule had more room than in its initial position, neatly arrayed with the others. It would then start to rotate like a spinning top with a diameter of 1.2 nanometers.

The energy generating the spin came from the thermal energy of the surface, which was maintained at room temperature. This experiment brought us the first image of a single spinning molecule. Once over the first excitement, we refined the experiment and found the parameters that control the rotation.

After several weeks of research, Gimzewski and his colleague Reto Schlittler showed that it was possible to use the tip to start and stop the rotation of the molecule at will. We also explained the physics of this phenomenon. In fact, this molecular wheel works a bit like a gear. When it is at the edge of the space, four of its six legs mesh with the legs of the neighboring molecules and the molecule does not spin. A 0.25-nanometer shove moves it into the empty part of the interstice, releasing its four legs and allowing it to rotate. The space needs to be big enough for it to be able to spin, but not too big, otherwise a process of lateral diffusion is superimposed on the rotation.

In order to understand how this molecule rotates, we recorded the variations in the tunneling current when the tip was positioned at the precise spot through which one of its legs moved when the molecule was spinning. We expected to see the current oscillate on the oscilloscope screen at the same rate as the leg passed under the tip. Unfortunately, at room temperature, our molecular wheel spun too fast and we could not get any recordings. With our colleagues at Berlin Free University, we therefore designed and synthesized another molecule, this time with six longer teeth, more like a real cogwheel with a diameter of 1.2 nanometers. We chemically marked one of the six teeth by slightly modifying its composition and were able to observe the rotation of the molecule: it turned step by step, at will, in 60° segments, within a rack also made up of molecules.

In 2001, working with Francesca Moresco and Gerhard Meyer, we repeated the experiment of pushing the four-legged phthalocyanine molecule with the STM tip, and made a real-time recording of the current variations between the tip and the surface. On the oscilloscope, we observed large regular oscillations with a period of 0.25 nanometers, which showed that the molecule was moving from site to site on the surface of the copper. Within these large oscillations, there were also smaller oscillations, corresponding to the alternating movements of the two "front" legs — those facing in the direction of movement (the "rear" legs were blocked by the tip). When the molecule was pushed, it was as if it were crawling along the surface, alternately changing the shape of one leg, then the other. This deformation slightly alters the molecule's electronic structure. As a result, the current that travels through the junction formed by the surface, the molecule and the tip varies at the same pace as the movement of the front legs. In order to get to the next site, therefore, the molecule moves one leg after the other, rather than the two legs together, as if it were walking.

All the experiments described so far can be understood by well-known laws of physics. Here, however, is one that cannot yet be explained. Don Eigler continued his experiments with xenon atoms on the surface of a metal. He ran a "strong" electric current through an atom. What happened? On our scale, when an electric current passes through a sample of matter, the sample heats up. This is the Joule effect, which is exploited in electric radiators. However, when a strong current passes through a xenon atom, Eigler observed, the atom jumps on the tip, or further away on the surface. The probability of the jump happening depends on the intensity of the current. But whereas with the Joule effect, on our scale, the power dissipated in a material varies by the square of the intensity of the current, the probability of the atom jumping varies here to the fifth power of the intensity of the current. So far, no one has found an explanation for this difference, or for the source of this fifth power. Wilson Ho, of Irvine University in California, has studied a similar question. He looked at the probability of rotation occurring in a small acetylene molecule on a metal surface with the application of different levels of current. As the intensity of the current increases, it begins to rotate and to jump from site to site, with a probability that also depends on the intensity of the tunnel current applied. Eigler and Ho had shown that the physical laws which the molecules follow here have no equivalent in the macroscopic world or on the mesoscopic scale ....

The Advantage of Staying at the Bottom

Many more nanophysics experiments on a single atom or a single molecule have been conducted since the early 1990s. They provide direct access to the world at the bottom, allowing us to study physical phenomena that can be observed with the minimum possible quantity of matter. In these experiments, a single atom or a single molecule is enough. The essential difficulty is to adapt — often even to reinvent — the appropriate measuring instrument. So the aim is no longer to adapt the sample to an existing measurement technique but, vice versa, to adjust the measuring instrument to the size of the object to be measured. In other words, a measuring instrument measures another device smaller than itself, and so on, like a Russian doll, right down to the scale of the individual molecule.

These experiments are creating a new field of knowledge and a new scientific program. The first goal of this program is to encourage the building of experimental devices constructed atom by atom or with a single molecule. They have all the power of their macroscopic equivalents. Some have even become "epistemological" devices, in that they are prompting us to re-examine the laws of physics as we understand them today. We will give some examples in the next chapter.

The second goal is more fundamental. By constructing devices at the quantum scale, physicists hope to lift a corner of the veil that still hides the quantum realm. In these early years of the 21st century, we are in a position to test the edifice of quantum mechanics in a new way. Will this lead to the birth of a new science? Will new laws emerge from the atom-by-atom manipulation of matter? Over the centuries, a host of new experimental practices have been developed, but very few new sciences have been born. If a new phenomenon were to emerge from the exploration of the world at the bottom, which could not be explained by the laws of quantum physics, then a new science would come into being — science at the nanometric scale, or nanoscience. Otherwise, what would be the point of inventing a new word to define the field of technical knowledge opened up by nanotechnology?

CHAPTER 4

MONUMENTALIZATION

By making it possible to build devices that work with no more than a single molecule or a few atoms, nanotechnology is reversing the standard approach in a way that runs counter to our ancestral pursuit of technological miniaturization. These minuscule devices fascinate scientists seeking to understand the nanophysics that underpins them. Might it be possible to go further, to open up new paths, by enlarging the molecule, increasing the number of atoms it contains, transforming it, for example, into a calculating machine or a mechanical device? That would end the problems of miniaturization in microelectronics or micromechanics: everything would be in the molecule and the molecule would become the machine. That is why these monumental structures have been christened molecule machines — molecules that will become increasingly monumental as the complexity of the machine that they embody rises.

To begin the process of "monumentalization," we first have to determine the number of atoms required for a molecule machine such as an engine, a two-way transmitter or a computer, to work. We then have to design the machine, equipping it with parts that maintain its structural stability and others that perform functions. And, finally, we have to devise the technical means of giving it instructions, supplying it with energy or allowing an exchange of information.

This idea of monumentalization emerged in the early 1980s, when Forrest Carter, a chemist at the NRL (Naval Research Laboratory), had the first inklings of the concept. He was working on polymer conductors, large thread-shaped molecules which, when grouped in a material, form electricity-conducting plastics. While studying these molecules, Carter realized that the molecular electronics imagined by Ari Aviram, in which each component of an electronic circuit would consist of a single molecule, would encounter the same technological barrier as transistor electronics at the end of the 1950s. At that time, electronic circuits were assembled component by component, and it seemed unlikely that it would one day be possible to connect the millions of components required to build a computer. Jack Kilby resolved this problem for solid state electronics by inventing the integrated circuit in 1958.

Carter realized that it would be equally impossible to assemble millions of molecular components to form an electronic circuit, especially as the interconnections in such a circuit would still be metal wires, and each component would have to be separated from its neighbor by at least 10 nanometers or so, otherwise the operation of the circuit would be disrupted by quantum effects. In these circumstances, the interconnections would take up considerable space, so what would be the point of trying to convert a molecule into a component if the resulting circuit was no smaller?

Just as Kilby had resolved the problem of connecting electronic components by inventing the integrated circuit, in 1984 Carter proposed a solution to the problem of connecting molecular components by devising an "integrated molecular circuit," consisting not of components reduced to single molecules, but whole circuits embodied in a single molecule. In other words, he moved from the idea of the molecule as component to that of the molecule as circuit — a single molecule performing as an entire electronic circuit. He then began to design monumental molecules which could operate as such circuits — in other words, which would incorporate both the electronic components and the interconnecting wires. There was outrage in the physics and chemistry fraternities! At the time, there was little acceptance that a single molecule could even become an electronic component, let alone an entire circuit in its own right.

Nonetheless, Carter got backing from a few biotechnologists in California, including Kevin Ulmer of the firm Genex, who could already see themselves genetically programming bacteria to manufacture, not proteins, but molecular electronic circuits directly on demand. In France, the firms Roussel-Uclaf and Elf Aquitaine, together with the Institut Pasteur, were ready to come on board. They had sent emissaries, including the then director of research at the Institut Pasteur, Joël de Rosnay,[11] to the first Molecular Electronics Conference, organized by Carter, before deciding

---

[11] Joël de Rosnay, "Les biotransistors: la microélectronique du XXIe siècle," *La Recherche*, July–August 1981.

that the research was too risky. It is true that connecting the molecule to contact plates in order to link it to the macroscopic world and exchange information or energy with it, seemed an insoluble problem. Nevertheless, at the end of the 1980s, Aviram got behind the idea of molecule circuits and diverted part of his research into this new field of molecular electronics.

At the same time, Eric Drexler was beginning to design complex molecular machines, based on concepts such as molecular gears. He designed several examples on the computer, incorporating advances in molecular modeling techniques. However, this stage of monumentalization developed independently of chemists and these molecule machines remained virtual. Later on, with the discovery of the STM (with its ability to manipulate molecules), chemists reworked these computer-designed molecules. They reduced their complexity in order to find ways of synthesizing them and giving them reality.

The Advent of Molecule Devices

Having looked at the emergence of the concept of the monumental molecule and the first virtual molecule machines, let us move on now to the first molecule devices actually built. These were not as yet calculating machines, but devices already capable of making measurements at the molecular scale. Let us start by opening a mid-20th-century physics textbook. It shows ingenious experimental apparatuses designed to study ill-understood physical phenomena. They measure such things as the impact of temperature on the conductivity of a semiconductor or the amplification of an electric signal by means of a piece of metalized Plexiglas held at the surface of a semiconductor (the principle of the transistor). By reversing the technology of manufacture, nanotechnology has raised the possibility of writing a new textbook, in which each of the old devices is replaced by a single molecule, which becomes simultaneously the experimental apparatus and the subject of the experiment.

A Wire …

The first molecule device made for a physics experiment was a molecular wire fitted with four molecular legs. It was designed by André Gourdon, from the Centre d'élaboration de matériaux et d'études structurales (CEMES) in Toulouse, and myself in 1997. We nicknamed it *Lander*, because it reminded us of the little *Sojourner* robot on the *Mars Pathfinder* probe which NASA had just landed on Mars that summer. Gourdon succeeded in synthesizing it a short time later.

The aim of this experiment was to measure the electric conductance of the molecular wire. We had mounted it on four legs to elevate it above the metal surface, which would otherwise have created leakage currents. It was easy to manipulate a four-legged wire like this on a flat metal surface using the tip of the STM. The real difficulty was to establish the electric contacts at its two ends. To resolve this problem, we came up with the idea of exploiting a property arising from the preparation of the metal surfaces, which involves a process of alternately annealing and pickling that generates large flat terraces. By applying the right preparation temperature, it was possible to form terraces with dimensions in the hundreds of nanometers, which ended with a step the height of a layer of atoms that was easy to spot with an STM. The idea was then to use the tip of the microscope to position the molecular wire perpendicular to a step, then push it until its extremity was above that step. The extremity would then interact with the top of the step and bend, establishing the desired electric contact. The second connection was made with the tip of the STM, which had to be positioned exactly at the opposite extremity of the molecular wire.

In this experiment, the metal surface was the test bench, the molecule was the experimental apparatus used to position the molecular wire so that its resistance could be measured, and the tip of the microscope became an extension of the physicist's finger. It was Jim Gimzewski who, in his IBM research lab near Zürich, manipulated the *Lander* for the first time in 1998. He succeeded, as planned, in connecting it to a step, and in measuring the electric resistance of the contact between the step and the end of the molecular wire. This resistance was very high, much too high for electricity to flow well through the wire. The contact was poor, because the legs held the wire too high above the step, and because the chemical group at the end of the molecular wire was not extensive enough to provide good electronic interaction. By altering its chemical composition, we managed to reduce the contact's resistance tenfold. We intend to reduce it further by shortening the length of the molecule, and are looking for solutions to establish the contact at the other end of the wire, which would allow its resistance to be measured accurately.

An Ampermeter …

After this, we designed a more complex molecule device, in this case a molecule ampermeter, i.e. a molecule capable of measuring the intensity of a current flowing, for example, through a molecular wire connected to a metal electrode at each end. In this device, an electric current flows through the main branch, into which a small rotating chemical group has been inserted. When an electron is transferred from one electrode to the other through the molecule, a minute quantity of energy is always dissipated into the molecular wire. This is enough to heat the rotating group gradually and to modify its orientation. The angle of rotation can be determined by means of a third electrode positioned on the side indicating the rotating group, then by measuring the tunneling current. From the angle of rotation, the experimenter can measure the intensity of the electric current flowing through the main molecular branch.

This molecule ampermeter must be fitted with legs to elevate it above the surface and to allow the small rotating group to spin. This means that it requires three electric contacts. The ideal would be to make two metal contact plates with atomic precision on the surface of the solid and to push the molecule until it connects with the plates (the third contact being held by the tip of the microscope), but that technology does not yet exist. It is the subject of intense research and should soon be available, since it is also crucial to the production of "calculating molecules" — molecules capable of adding or subtracting numbers, as we will see later in this chapter.

The first transistor, made at the end of the 1940s, had shown that a solid state device could amplify an electric signal. Fifty years on, following the pioneering work done by Ari Aviram and Mark Ratner, the question raised by molecular electronics was whether a single molecule could also amplify an electric signal. So researchers interested in the field were testing the resources of molecular structures capable of achieving this. In 1997, Gimzewski and I answered this question by showing that a single molecule can indeed amplify an electric signal. This was not going to knock microelectronics off its throne, but for us it represented a considerable advance.

We put together an electric setup in which a fullerene molecule is placed under the tip of an STM. We knew that, when the tip squashes the molecule slightly, its resistance rapidly decreases. A small variation in one parameter (here, the distance between the tip and the surface) produces a large variation in another parameter (in this case, the molecule's resistance). We used this effect to produce an amplifying device in which the output voltage is four times higher than the input voltage.

Our next idea was to link a large number of these devices in series and parallel to form a proper electronic circuit, capable, for example, of performing calculations. Unfortunately, the electric resistors required for each molecular amplifier to work, and the interconnecting wires, were not molecular. In fact, the interconnections were macroscopic and were situated outside the structure containing the STM. We tried to miniaturize them to fit inside, but it was a waste of time. So, instead of trying to shrink the apparatus around the molecule, we changed tack and gave up the idea of hybrid molecular electronics, of combining molecular and microscopic components. Instead, we wondered if it might not be possible to "fatten up" the molecule so that it could contain all the resistors and connecting wires needed to constitute a full electric circuit. This was the beginning of "monumentalization."

This break with the historical idea of hybrid molecular electronics brought us within the ambit of Forrest Carter's work. It was something of a wrench for me to quit the field of molecular electronics, which I had been working on since the late 1970s, a matter of almost twenty years. However, others have continued along this path and are looking at a number of small molecules, which, once deposited on the surface of a metal or a semiconductor, display several orientations (with almost equivalent energy). They constitute "natural" switches. Indeed, with the tip of an STM, it is easy to shift from one orientation to the next, and thereby to produce a switch with a single molecule. But however attractive these molecular components are, assembling them into a circuit raises the problem we encountered with our molecule amplifier. We now focused on the new question that clearly confronted us: How much computing power does a single molecule have, compared with a circuit made up of a myriad of assembled molecules?

This idea of a new molecular electronics totally integrated within a single molecule has released researchers from the assumption that only a small device could be made with a molecule, at best. It has opened their horizons by allowing them to entertain the possibility that a molecule might fulfill much more complicated functions. In the remainder of this chapter, we will see that this shift concealed another, more profound shift, in quantum physics itself. For my part, I immersed myself in increasingly complex molecule devices, like the Morse molecule.

And a Cantilever

A Morse manipulator is a little device familiar to lovers of westerns: the cap-wearing operator frenetically taps away at a Morse code key, trying to send a coded message to the neighboring station over the telegraph wires. In practice, a Morse manipulator consists of a small metal cantilever, with a button at one end and a return spring at the other. This allows the operator to close an electric circuit at will and to send a coded signal made up of a succession of long and short pulses.

A molecule has recently been designed to perform the function of a Morse manipulator. The cantilever is a molecular arm held parallel to the surface. One end is fixed by a chemical bond (acting as the return spring) to a four-legged molecule which acts as the central board. The other end is suspended; it is this end that the operator presses with the tip of the STM.

This molecule, less than 1.5 nanometers long, is one of the most complex molecule devices so far designed, and it could take several years to synthesize chemically. It will only be able to function on a metal surface; when the small chemical grouping at the free end of the cantilever approaches the surface, it modifies its local electron state and alters its electronic density, which can be detected a little further along the surface.

It is not only molecules that are used as an experimental apparatus: experiments have also been carried out with atoms arranged on a surface. Large numbers may be involved, but no more than are strictly required for the experimental apparatus. For example, Don Eigler ran an experiment in atomic magnetism in an elliptical enclosure built with 36 cobalt atoms, individually manipulated with the tip of his STM on the surface of a copper crystal. These atoms have the property of reflecting the quantum waves associated with the electrons that circulate freely on the surface of the copper. Their 1.5-nanometer wavelength allows us to observe their interferences at the center of the atomic enclosure, which is a few nanometers long. With his STM, Eigler has produced images of these interferences, which show a succession of concentric circles. These pictures have been seen all around the world. They provide a spectacular demonstration of the wave nature of electron states on the surface of a metal. Eigler continued his experiment by placing a magnetic atom with the tip of the microscope at one of the focal points of this atomic oval. He observed a magnetic echo … at the other focal point, although there were no atoms on that focal point. This was a magnificent magnetic mirage effect, transmitted from one focal point of the ellipse to the other by means of the electron cloud on the metal surface. Of course, it is possible to carry out this experiment on all scales, with light or sound waves. All you have to do is choose an elliptically shaped resonator with the right dimensions for the wavelengths used.

Molecule Machines

Let us move on to mechanics. Molecules designed for use as mechanical machines must carry on board all the parts necessary for them to work. This means that they are more complex than the molecular devices we have just described, since they necessarily have different mechanical parts (which, at 1.2-nanometer-long have to be held by strong chemical bonds). In 2001, we designed a machine which we dubbed a "molecular wheelbarrow." It had two molecular front wheels, with a diameter of 0.7 nanometers, attached to an axle; two legs at the rear, like the legs of a wheelbarrow; and finally two little sleeves at the back to act as handles, where the tip of the STM would push. These wheelbarrows were synthesized by Gwenaël Rapenne at CEMES in Toulouse. Then, Leonhard Grill and Francesca Moresco, of Berlin Free University, were responsible for launching and manipulating the wheelbarrows. In the launch phase, the molecules — generally prepared in powder form in a small crucible — were usually vaporized by heating the crucible to between 150 and 250°C and positioning it in such a way that some of the molecules were deposited on the surface. However, to vaporize large molecules like molecular wheelbarrows, the powder had to be heated to between 350 and 450°C. In these conditions, almost 95% of the wheel-barrows were destroyed as they left the crucible or reached the surface. Amongst the molecular residues on the surface, new molecules, with two, three or four wheels, appeared — some linked by axles much shorter than the original — which might be described as broken or reassembled wheelbarrows. At high temperatures, therefore, wheelbarrow fragments spread randomly across the surface and reacted to form new, small-scale molecules.

Fortunately, many of the molecular wheelbarrows reached the surface intact. We tried to push one of them from behind using the tip of the STM, in order to persuade its two front wheels to turn. No joy! After several attempts, we observed that the legs were bending, but the two front wheels were not turning. Later, we understood that the wheels interacted too strongly with the metal surface. This failure illustrates the difficulty we have, when designing molecular nanomachines, in escaping concepts drawn from the structure of machines in our macroscopic world.

During this time, James Tour, a professor of chemistry at Rice University in Texas, was working on the creation of a molecular car. He designed a nanocar 1.5 nanometers long, with four wheels, each made of a single fullerene molecule. Each wheel was linked to a small molecular axle identical to the one on the wheelbarrow. This molecular car could be manipulated with the tip of an STM, but the easiest way to make it move forward was to heat the gold surface on which it was deposited, so that the thermal energy of the surface would move it spontaneously and randomly around the surface from atomic site to atomic site. Here, however, the wheels did not turn either. Because the intensity of the tunneling current captured by the tip of the microscope was sensitive to variations in the internal shape of the molecule, the tunneling current should have notified the operator whether the wheels were rotating. Instead of turning, however, they appeared to slide across the surface. It would seem that the surface of the fullerene wheels was too smooth.

Despite these difficulties, progress is being made and nobody doubts that working nanovehicles are on the way. However, they will need to become more autonomous and, for example, have their own engine. James Tour has already incorporated a small ratchet into the center of his molecular automobile's chassis. The idea is that, activated by light, this ratchet will move close to the surface and grip it to move the molecule forward. This molecular ratchet car is in the process of being synthesized. At the Toulouse laboratory, Gwenaël Rapenne and Jean-Pierre Launay have designed and synthesized a molecular motor with a drive wheel less than 2 nanometers in diameter. They are in the process of assessing its driving power and devising a transmission belt so that it can be incorporated into a molecular automobile.

Calculating Molecules

As long ago as 1997, Jim Gimzewski built a little molecular counting frame by assembling ten fullerene molecules, one by one, along a step one atom deep naturally present on the surface of a gold crystal. In 2002, Don Eigler manipulated a hundred or so carbon monoxide molecules, one by one, with the tip of his STM, to build logic gates that would perform "OR" and "AND" operations. These logic gates have two inputs (through which the signal 0 or 1 arrives) and one output. If one of the two inputs is on 1, then the "OR" function positions its output to 1, whereas the "AND" sets its output to 1 only if both inputs are on 1. Eigler aligned the molecules in such a way that they formed two rows converging at the same point on the surface, to form the two inputs to the molecular logic gate. Each molecule is like a domino (0 or 1). When the first molecule flips, the other molecules in the row also flip, as in the domino effect. Depending on the flip state of the last molecule in the row before the gate (0 or 1), the output either "flips" or stays the same. That is how a set of molecules can form a logic gate. However, this gate works only once. To make a new calculation, all the molecules have to be reset, like lining up all the dominoes again.

It is therefore possible to perform a mechanical or electronic calculation with a set of molecules. However, although it is possible to synthesize molecular logic gates, it is as yet impossible to make them work, because the technology for connecting them to contact plates does not exist yet.

In the 1990s, it was thought that electronic lithography would allow us to manufacture metal electrodes small enough to converge on a single molecule. This proved impossible. Firstly, electronic lithography on the mesoscopic scale is not precise enough to control the ends of these atomic scale electrodes. Secondly, this technique is dependent on the use of a resin (to transfer the pattern of the metal contact plates on to the surface), which must be cleaned down to the last molecule after use — with the risk of cleaning away the molecules in the device itself. All hopes rest on a technique of depositing small metal clusters a few nanometers long, which can be manipulated with the tip of the microscope to build a set of nanocontact plates.

While work continues on this new technology, researchers are designing new architectures for molecule computers. Molecules of the kind suggested by Forrest Carter are gigantic, and will become even larger as the complexity of the calculation to be incorporated into the molecule increases.

This poses several problems. First, large molecules are hard to synthesize. Second, it is also difficult to manipulate gigantic flat molecules individually with sufficient precision to connect them to metal contact plates. And, finally, the electric current that can flow through a very long molecule is undoubtedly less than an attoampere, i.e. one billionth of a billionth of an amp. Fast electronics cannot be developed with such low currents. New architectures for calculating molecules are in the process of emerging from the alliance between molecular electronics and quantum computers.

Quantum Computing Molecules

The concept of the quantum computer was invented in the 1980s by Richard Feynman and David Deutsch, from Oxford University's Centre for Quantum Computation. The principle is to exploit the spontaneous response of a quantum atomic or molecular system, in a nonstationary state, to perform calculations. The system must be divided into small computational units, called "quantum bits," which can exist in a quantum superposition of two states (0 or 1) and interact without exchanging electrons. The calculation is then performed by simply allowing the state of all the quantum bits to evolve over time. A quantum computer is a little like those ball clocks where ball bearings roll along rails of different lengths, telling the time as effectively as a clockwork timepiece. With a quantum bit system, the initial state of the quantum bits encodes two numbers which have to be added together. The system then evolves over time until it reaches another quantum state, which gives the result of the calculation.

This quantum computer concept shows that you do not need an electronic circuit to make a computer. Back in the field of molecular electronics, this shows that there is no need to "force" a molecule to look like an electronic circuit so as to give it a computational function. We can simply use the intrinsic quantum dynamics of a molecule to perform a calculation. Molecular quantum computers can perform all sorts of operations and, for equivalent complexity, will be smaller than the molecular circuits conceived by Forrest Carter. Researchers have even recently demonstrated that it is not necessary to divide the molecule into quantum bits to perform quantum calculations. One can simply manipulate the electronic structure in order to modify the molecule's inherent quantum development over time. The molecules are currently being synthesized and will soon enter the experimental phase. They would provide a way out of Moore's famous law. With quantum computing, increasing the complexity of a computer would no longer be dependent on fitting an ever-greater number of transistors on the surface of a semiconductor, but on the ability to control the temporal development of a quantum system that grows increasingly complex with each generation in the quantum state space.

Molecular Factories

We have described the first molecular machines and the first molecular computers. It is tempting to combine the two types of molecular devices and to place a molecular computer on board a molecular machine in order to make ... a molecular robot. In our macroscopic world, a robot is essentially a set of mechanical functions controlled by an on-board computer. For the moment, molecular robots are just an idea. No one knows if they can be made. It seems likely that both the chemical synthesis and the remote control of such nanorobots would be very difficult.

Since synthesizing nanorobots is a problem, researchers have proposed delegating the task to other molecular machines. Their role would be to assemble molecular machines of all kinds atom by atom (or chemical group by chemical group). They do not specify how these molecular assemblers — production plants for molecule computers, molecule machines or nanorobots — would themselves be made. In our current state of knowledge, such a task is of course unrealistic.

As described, these molecular assemblers would be exact ultra-miniaturized equivalents of the robots used in our factories. For example, they would have grabs and telescopic arms for grasping and assembling molecules one by one. Richard Smalley, codiscoverer of fullerene molecules, has objected that, if a molecular grab grasped a molecule, it would never be able to let go of it, since the chemical reaction required to capture it would simply prevent it being released. In fact, there is no evidence that an assembler needs a grab at the end of a telescopic arm to grasp an atom or a molecule. For example, by manipulating a small six-legged molecule with the tip of an STM, our colleagues from Berlin Free University have discovered that it can "vacuum up" a few copper atoms previously deposited on the surface. These atoms accumulate one after the other under the molecule, and cannot re-emerge because they are trapped by the molecule's legs. The experimenter can release the atoms when he wants by using the tip of the microscope to remove the molecule that assembled them.

Other researchers, like Wilson Ho, Don Eigler, Gerhard Meyer or Gérald Dujardin of Paris-Sud University at Orsay in France, are experimenting with the use of the STM as an assembler. They are trying to use it to synthesize molecules atom by atom, or molecular fragment by molecular fragment. However, it is extremely difficult to persuade two molecules to react chemically using the tip of the microscope. They have to be manipulated precisely to achieve the orientation that will allow the chemical reaction to occur. In solution, the problem does not arise, since thermal agitation causes the molecules to explore multiple random orientations and therefore to achieve the chemical reaction spontaneously.

Bigger and Bigger?

Monumentalize, sure, but how and by how much? Up to a certain size, we will be able to synthesize a molecular machine in a single operation. In 2002, for example, Japanese chemists succeeded in producing a molecular wire 100 nanometers long. However, beyond a certain size or a certain complexity (but which?), it will no longer be possible to synthesize a molecule composed of ever-more-complicated mechanisms in a single operation. We will need to go back to the standard production line concept. This brings us into the field of supramolecular chemistry, already well covered by chemists like Fraser Stoddart at the University of Los Angeles, Jean-Pierre Sauvage at the CNRS in Strasbourg, and Nobel Prize-winner Jean-Marie Lehn, of the University of Strasbourg.

How do you assemble a large number of molecular parts to build a complex machine? Some scientists, like Lehn, are studying spontaneous assembly, where the elementary parts are marked, like the pieces of a puzzle, with specific chemical groups. Each chemical group recognizes its matching group, belonging to another piece of the puzzle, until all the groups fit together. This method of spontaneous assembly happens in viruses and in living creatures like bacteria. Hence the importance of studying viruses, which might be classed as the simplest self-assembled machines or "automats."

In 2002, the team headed by Professor Eckard Wimmer at Stony Brook University near New York was the first to synthesize a virus — the poliovirus. In the wild state, this virus takes the form of a little ball 28 nanometers in diameter. Its protein and gene structure was discovered in 2000 by James Hogle of Harvard University, then in 2001 by Wimmer's team. The virus has a viral part and a more or less spherical envelope. The viral part is an RNA macromolecule containing 7411 nucleotides which, once unraveled, are a few micrometers long. The envelope containing the viral part is composed of 60 subunits, each comprising 4 proteins. Each protein is, on average, made up of 250 amino acids. In 2002, Wimmer's team first synthesized the RNA of the viral part with a total length of 7411 nucleotides, most of the sequences of which can be bought from private biotechnology firms. Then it synthesized the missing bits chemically. Compared with the multistep chemical synthesis of a complex molecular machine, the task here is greatly simplified because the molecular pattern to be synthesized is regularly repeated. Once this viral part had been obtained, the researchers did not try to synthesize the four proteins that constitute the envelope. To obtain them, and above all to obtain them assembled in the right order, they used a "soup" of living cells, so that the synthetic polio RNA would use the cellular machinery to create and self-assemble the envelope. The idea of exploiting the natural chemical factories that bacteria represent could be extrapolated to make molecule machines.

The Retreat to Nanomaterials

We have just described the first molecule machines and possible ways of manufacturing them with sufficient size, around a dozen nanometers, to perform the complex functions usually carried out by the machines familiar to us on our scale. However, few researchers have taken on this challenge of monumentalization, compared with those who have made nanomaterials their research focus. The term "nanomaterials" refers to concretes, cladding, ceramics... tangible applications that are a long way from the field of molecular machinery. They do not belong to the nanometric scale, so what is "nano" about them? The term "nanomaterials" is an abbreviation of "materials structured on the nanometric scale." It refers to materials whose elementary structure consists of molecules, macromolecules or small clusters of atoms, which are nanometric in size. Let us look at our example of table salt (or sodium chloride). Its basic structure consists of a chlorine atom and a sodium atom (a little less than 0.3 nanometers apart). This spatial pattern is repeated to create a small crystal of salt with square sides, which can be picked up with tweezers and seen through a magnifying glass. Table salt is therefore an atomically structured material. The basic element of a nanomaterial is a molecule, which can be complex and give the nanomaterial a specific and very important property (e.g. resistance to bending, or capacity to record information). This property appears only when millions of identical molecules are assembled. The same is true of nanoparticles, which have a diameter of only a few nanometers, but are composed of an arrangement of thousands of atoms.

The effects of structuring matter on the basis of the properties of materials have been known and exploited for millennia. Even in ancient times, copper nanoparticles were incorporated into glass to redden it. Paints are solutions containing an emulsion of nanoparticles. The carbon black or smoke black described in 19th century manuals is made up of carbon nanoparticles with a diameter of 10–1000 nanometers, which were used and are still used as pigments in ink. These carbon black nanoparticles were introduced into tires in 1917 to increase their longevity. Catalytic converters on cars contain nanoparticles of platinum, rhodium and palladium, arranged in the minute pores of a ceramic block, in order to increase the surface area in contact with exhaust gases. They boost the chemical reactions which convert the carbon monoxide and nitrogen oxides in exhaust gases into water and carbon dioxide.

What distinguishes nanomaterials from traditional materials is the greater complexity of the chemical structure of the base pattern. The research field of nanomaterials has nothing to do with monumentalization, in which the molecule becomes a machine, whereas in a nanomaterial it remains an elementary building block. Nanomaterials constitute a huge field of research which would merit a separate book all to themselves. They are not, however, nanotechnology.

Many scientists continue to see the molecule as no more than a tiny "fragment" of matter. A few years ago, the idea of a molecule machine was even a source of some derision in scientific circles. Yet it is now an experimental reality. The status of the molecule has changed radically, from an anonymous presence within a multitude to an individual existence accessible to measuring instruments. The challenge of monumentalization today is to find new ways of manufacturing these molecular devices and machines, which will contain only the number of atoms they need to function. It would be a clever person who could predict the size and degree of complexity that these machines will attain.

CHAPTER 5

NANNOBACTERIA

From its status as an elementary building block lost amongst billions and billions of others, the molecule achieved its independence fifteen years ago. It can now embody a device or a machine capable of performing increasingly varied and complex functions, and monumental assemblages of these molecule machines will soon be built. Once we have understood the construction plans of the proteins, membranes and ribosomes operating in a living cell, will we be able to reproduce the architecture and organization of the smallest known forms of life? And, once these components have been assembled, will this artificial cell be alive?

The smallest known living organisms on our planet today are bacteria. And the smallest bacteria measure 200 nanometers, which is minute. In comparison, the most common bacteria often reach a size of 1000 nanometers and the cells of the human body measure around 20,000 nanometers. Viruses are much smaller (between 20 and 300 nanometers), but they are not considered autonomous living organisms, because they are incapable of reproducing independently.

However, there is a possibility that even smaller bacteria have been discovered, thought to measure as little as 100 nanometers, and in some cases even 20. If they exist, these "nannobacteria"[12] are smaller than anything previously recognized as living. They are so small that they simply should not exist. Indeed, to feed and reproduce — in other words, to be alive — an organism must in principle possess the essentials of life, in this case DNA, ribosomes to manufacture proteins and a cytoplasm, i.e. a gel to contain these components. All this is sealed within the plasma membrane, protected by a rigid barrier (though these are lacking in mycoplasms), all of which

---

[12]For an explanation of the double "n" spelling, see Appendix II.

requires space. Theorists have calculated that a living organism cannot be smaller than 180 nanometers.

Yet these new bacteria are apparently smaller. Depending on one's point of view, they are either the smallest assembly of macromolecules capable of life (top-down perspective — miniaturization), or a monumental natural macromolecular machine with the property of life (bottom-up perspective — monumentalization). Their existence has aroused controversy. Luc Montagnier, the discovery of the AIDS virus, has described them as UBOs: "unidentified bacterial objects." These (as yet) unidentified bacterial objects have been found in different places on the planet: in underwater rocks and sediments, but also in mammals (cows and human beings).

Ripples from a Meteorite

Until 1996, the diameter of the smallest known bacterium was larger than the theoretical limit. Everything was fine. True, there were the strange objects found by University of Mississippi geologist Robert Folk, around hot springs at Viterbo, near Rome, in 1990. These object were minuscule (25–200 nanometers), round or oval in shape, and arranged in chains or clusters. Since, according to him, these features had never been encountered in anything mineral, he claimed that it was a new form of life, which he christened "nannobacteria," with a double "n," employing the prefix "nanno" used in the early 20th century and perpetuated in biology and paleontology (see Appendix II). However, Folk was not taken seriously: what did a geologist know about biology? Life cannot exist at that scale! What Folk had discovered was probably merely the debris of common bacteria, or slightly unusual mineral accretions....

Nonetheless, Folk was adamant about his discovery. In 1992, he gave a lecture to the American Geology Society, and this time his ideas did not fall on deaf ears. One of the people in the audience was Chris Romanek, a NASA geochemist. He decided to look for such objects in his samples. He discussed it with David McKay, of NASA's Johnson Space Center in Houston, and they, armed with a high resolution electron microscope — and a stack of patience — looked for and identified similar structures to those found by Folk.

This discovery was to have a worldwide impact, since the samples analyzed by the researchers came from a somewhat surprising source — a

Martian meteorite. The fall of this meteorite, named ALH84001 (because it was discovered in 1984 at the foot of the Allan Hills mountain chain in Antarctica), had gone unnoticed, as had its discovery. But when, twelve years later, McKay and Romanek found in it structures similar in size and shape to the nannobacteria discovered by Folk, it triggered a major row. Not only did the discovery seem to confirm the existence of nannobacteria (which was big news in its own right), but it also suggested that these nannobacteria came from Mars, that there had once been life on Mars. The whole world held its breath. Exit little green men — extraterrestrials were apparently bacteria smaller than anything we had imagined. Some cried foul: this "discovery" came just a few weeks before Congress was to vote on funding for missions to Mars. A long debate ensued: Were these elongated shapes really traces of life? Did they really come from Mars?

A team of French scientists decided to look more closely at a meteorite that had fallen in the Desert of Tataouine in Tunisia in 1936, which was in the National Museum of Natural History in Paris, and at other fragments collected in the 1990s. The Tataouine meteorite is not from Mars, but it is similar in composition to ALH84001. Surprise: they also found sticklike structures a few dozen nanometers long in the meteorite fragments, but — further surprise — nothing on the parent fragment, collected a few hours after it fell. That would mean that the "nannobacteria" had developed on the fragments that had remained in the desert since the meteorite fell in 1936, but not on the parent fragment protected from contamination in the museum. So these "nannobacteria" were of terrestrial origin, and the shapes found on ALH84001 could also have been caused by contamination … and not by extraterrestrial life.

This did not alter the fact that those elongated shapes were incompatible with the operation of life as we know it today. Since then, similar shapes, measuring 50–500 nanometers in size, have been found in Australia, in samples taken 10,000 feet down in the ocean. The geologist who made the discovery, Philippa Uwins, called them "nanobes" (for "nano"-biological organisms, as "microbe" means "micro"-biological organism). She has continued her research and shown that these nanobes are composed of carbon, oxygen and nitrogen, the characteristic elements of life. She has also proven that they grow spontaneously in a culture at room temperature, that they have a membrane and, above all, that they react positively to three tests for the presence of DNA. "If nanobes are not biological organisms, it is difficult to suggest anything else that might be compatible with these observations," the geologist concludes. However, she has not published anything on this subject since, which is surprising given the significance of the discovery.

Surrounded by Nanoaliens

In his research, the microbiologist Olavi Kajander, from Kuopio University in Finland, would often prepare broths of cell cultures. One day, all his attempts failed — the cells kept on dying. To find out what was causing the problem, he analyzed the fetal bovine serum added to the culture medium as a nutrient. The serum was not contaminated, but proved to contain unknown organisms, between 50 and 200 nanometers in size. In his view, they were nannobacteria, which meant that such entities were confined to rocks but are also present in living organisms. The detractors, for their part, were not backing down. They claimed that these samples were contaminated by other bacteria or by organic waste, or else contained ordinary bacteria that had shrunk as a result of stress.

Karim Benzerara, of France's Institute of Mineralogy and Physics of Condensed Media (CNRS), tried to end the controversy by penetrating the mystery of nannobacteria. He was a member of the team that had studied the Tataouine meteorite, and pursued their observations with much more powerful instruments — a very powerful transmission electron microscope (TEM) and a synchrotron (particle accelerator). He showed that each stick found in the Tataouine meteorite is in fact a calcite crystal, not a viable micro-organism. This would suggest that these peculiar shapes are actually produced by purely mineral processes — processes that have even been reproduced in the lab. It looked like a fatal blow to the theory of geological nannobacteria (those discovered in meteorites, rocks and underwater sediment).

Next, Benzerara applied his expertise to the human origin "nannobacteria" discovered in vascular tissues. Synchrotron studies had shown groups of carbon atoms characteristic of proteins, together with nanocrystals of calcium phosphate. Might this be proof that the concretions found in vascular tissue came from a new form of life? In fact, the proteins might have been incorporated into the concretions by accident. Benzerara now looked at the possibility that these human nannobacteria might have purely mineral origins. He suggested that they might be caused by the nucleation and growth of calcium phosphate crystals controlled by proteins. In his view, these nannobacteria arise from an as yet ill-understood process of synthesis, which is both mineral and organic in nature, but probably has nothing to do with life.

If, on the other hand, he were to show that these mineralo-organic objects are really produced by living creatures, it would be a revolution. These new bacteria could be implicated in diseases whose causes remain obscure, such as arteriosclerosis, renal calculus or the psammoma bodies implicated in ovarian cancer. They could also play a role in the formation of bones, teeth and dental plaque. More generally, they could contribute to mineral precipitation mechanisms which up to now have been considered exclusively inorganic. Above all, if they exist, nannobacteria would represent a new form of life, probably different from the form familiar to us today. Perhaps they might be archaic bacteria or protobacteria, a sort of missing link between molecules and the living bacteria we know?

The Missing Link

According to the most likely hypothesis, life emerged as part of a continuum running from the inert to the living, with molecules becoming increasingly complex and organized. The question in this case is: At what stage can we say that a given something is alive? It is a question that has taxed scientists since antiquity. In the fifth century BC, Greek philosophers believed that every grain of matter was alive. Lucretius thought that life was the result of the "grains" of the soul mixing with those of the body. In the 18th century, molecules were thought to be the smallest living beings in existence. Buffon (1707–1788) gave the name "organic molecules" to the first living cells observed after the invention of the optical microscope. How could you tell what was alive and what was not? At that time, this was far from clear. When, in 1827, the botanist Robert Brown observed through his microscope the continuous and erratic motion of grains of pollen dispersed in water (Brownian motion), he thought that he had discovered the "primitive molecule" responsible for life.[13]

---

[13] Henk Kubbinga, *L'Histoire du concept de molécule (op. cit.)*.

What made life particularly mysterious was that it seemed to emerge out of nothing. "There exists a tree ... frequently observed in Scotland. Leaves fall from this tree; on one side they fall in the water and are transformed into fish, on the other side they fall on the earth and are transformed into birds,"[14] claims a 17th century treatise on botany. This theory of spontaneous generation existed in different variants and only ended with the discoveries of Pasteur. Even then, he had to contend with diatribes from the theory's adherents for years, only managing to bring the dispute to an end in a lecture at the Sorbonne, where he showed that what looked like spontaneous generation was the result of microbial contamination and did not survive sterilization.

Similarly, vitalism, the belief that life is engendered by a particular force — one different from those that control physiochemical phenomena — persisted for a long time. It was disproved in 1828 when, for the first time, a living substance, urea, was synthesized in the laboratory by the German chemist Friedrich Wöhler, using standard physiochemical processes, without the need for any so-called "vital force." Since then, as our understanding of the chemical reactions that take place within the cell has improved, it has become apparent that the living world is governed by the same rules as the inanimate world. "Life is a product of molecular organization," asserts François Jacob, the French biologist who won the Nobel Prize in 1965.

Since Pasteur, we have known that life comes from life and, since Darwin, that species derived from each other. This means that all of us — human beings, carrots or snails — originate from a primary protobacterium ... or perhaps, if it should prove to exist, from a nannobacterium! One idea that is gaining credence is that the emergence of life is the outcome of chemical processes. Could this be reproduced in the lab? It goes without saying that this is too fascinating a riddle for scientists to leave alone.

The Molecular Fabrication of Life

The 20th century was the century of the manipulation of atoms. The 21st century will be that of the "manipulation" of life. The goal of "synthetic"

---

[14]Didier Pol, *Une petite histoire des recherches scientifiques sur l'origine de la vie*, INRP, http://www.inrp.fr/Acces/biotic/evol/orivie/html/histoire.htm.

biology, still in its infancy, is to recreate life. The scientists working in this field make no bones about it. But what is life? It would be useful to have a definition, just in case it should emerge one day on a laboratory test bench. It is tricky to define. The broad consensus is that life is the capacity for self-assembly into organized self-reproducing structures.

La Biologie synthétique (*Synthetic Biology*) is the title of a 1912 book by the French physician Stéphane Leduc, who was interested in the manifestations of life and tried to reproduce them in his laboratory. Using metallic salts in solutions of carbonate, phosphate or sodium silicate, he grew magnificent structures resembling waving seaweed, which appeared to be alive. The term "synthetic biology" next appeared in 1978, in an editorial in the journal *Gene* proclaiming the advent of the "era of synthetic biology, when biologists would no longer be content to describe existing genes, but would also seek to build new ones."[15]

Believing that life can be reduced to arrangements of complex molecules that form biological systems, the partisans of synthetic biology have no doubt that they will one day be able to reconstruct these systems from simple molecules. So they decipher the functions of the cell to understand how information flows, how systems of regulation are established, how genes and proteins interact, how a cell communicates with its neighbor, etc., and then try to reproduce these mechanisms. They also try to introduce new functions and to "program" the cell for new tasks. For example, a bacterium has been modified to emit green fluorescent light when certain molecules are present in its environment — a feat of which it was incapable, until it crossed paths with researchers.[16] There is no life without information, whether carried or transmitted. In most organisms, the information consists of sequences of DNA (deoxyribonucleic acid). That has not always been the case. Some biologists believe that RNA (ribonucleic acid) was previously the information carrier (as it still is in most viruses). In fact, DNA and RNA are similar, but RNA is also capable of accelerating chemical reactions millions of times, like an enzyme. It seems that RNA, as both information carrier and catalyst, preceded DNA, which is more stable and specialized in the transmission of information.

---

[15]Wacław Szybalski, in *Gene*, Vol. 4, No. 3, 1978, p. 181.

[16]Ron Weiss, University of Princeton.

What is this information used for? Put simply, DNA contains a comprehensive description of the cell. Its sequences are mission instructions, which are picked up by macromolecules that attach to DNA. They carry that information to "manufacturing machines" that make all the proteins needed for the survival of the cell. All the information is contained in just four letters, the nucleotides A, C, G and T, called bases. Researchers in synthetic biology can synthesize these bases and arrange them to manufacture artificial strands of DNA, which they study to see if they work like natural strands. They insert them into bacteria and observe what happens, in the hope of understanding life by imitating it and one day, perhaps, recreating it. Artificial strands of more than 100,000 bases have been produced, as well as more than 32,000 bases coding for certain proteins of the bacterium *Escherichia coli*. Others have gone further by manufacturing artificial bases other than the four bases of the living world (A, C, G and T). In 2002, for example, Japanese researchers created a six-base DNA, adding two artificial letters, S and Y, to the four standard bases A, C, G and T. Experiments have shown that bacteria are able to integrate unfamiliar bases. Might it be possible, by extending the genetic code in this way, that proteins that do not exist in nature, or new functions, or even other forms of life, could emerge? For the moment, no alien rabbit has yet been produced from the magician's hat.

Researchers have already synthesized the polio virus. After viruses, will the next step be to synthesize bacteria? The "manufacturing" principles might be the same. However, the genome of the very common bacterium *E. coli* contains 4.7 million bases! It is a task on a very different scale from the synthesis of the polio virus, which has "only" 7200 bases. So biologists are looking at simpler and smaller bacteria than *E. coli*. For example, the American biologist Craig Venter, who was part of the project to decode the human genome, is now working on the minuscule bacterium *Mycoplasma genitalium*, which has 517 genes, though this still represents 580,000 bases. Some genes seem to have no functions, but the question is: Which ones and how do you assess the minimum number of genes required for life? Biologists estimate the number at around 250, which does not seem out of reach. But this does not mean that life will be there once the necessary bases have been laid end to end.

Life is not that simple. Even if the genetic program can be built, you also have to manufacture the "box" (the container). Partisans of synthetic biology already have some ideas and have made a few attempts. For example, they have developed artificial vesicles capable of self-division when subjected to an external mechanical stress. They have also synthesized proteins capable of combining in a membrane to create a channel between the inside and the outside, through which nutrients and metabolic waste could flow. Rapid progress is being made with both container and content, but so far no one has managed to recreate life in the test tube. Could nanotechnologies contribute?

The Lessons of Mother Nature

Synthetic biology uses not nanotechnologies, but genetic engineering techniques. However, the new nanotechnology tools are a valuable aid in the understanding of the intimate functions of cells. Thus, the atomic force microscope, a derivative of the STM, can be used to "peel" back a cell membrane to look inside. Nanoprobes made with nanomaterials can be fitted to proteins and small viruses to track their pathways through cells. The confocal optical microscope can track these fluorescent markers *in vivo*.

Conversely, biology is sometimes useful to nanotechnologies: the works of Mother Nature have inspired the manufacture of certain nano-objects. For example, the study of the macromolecular sites (locks) situated on the surface of cellular membranes capable of recognizing proteins (keys), provides a model for constructing molecular "lock and key" devices. Nanoparticles fitted with "keys" could attach themselves to specific sites on diseased cells in order to deliver a drug. In a similar vein, the methods of self-assembly, scar formation or regeneration encountered in nature are being actively studied in order to develop nanomaterials that would assemble or repair themselves.

A close understanding of cell biology could also help scientists in the effort to manufacture increasingly complex molecules. They could fulfill the old, 1980s dream of using the biochemical factory — which is what a bacterium essentially is — to synthesize all or part of a molecular machine. This extreme monumentalization, bred in the very heart of a bacterium, is not a quest for artificial life, as the construction plans will be different from the macromolecules and organelles encountered in the bacterium, even if the molecular weight might be equivalent.

If it were to become possible to build increasingly complex molecular machines, might we see something living emerge as the byproduct of an experiment? Even if artificial molecular machines were equipped with the elementary functions present in a nannobacterium, does that mean they would be alive? Where is the breath of life in the millions of macromolecules that constitute a cell? The answer to this question has evaded scientific investigation, despite centuries of searching. The most modern hypotheses refer to the mechanisms of recognition and self-organization in molecules, but they are still a long way from explaining the nature of that spark of life that distinguishes the smallest bacterium from the most complex assembly of atoms and molecules that could be produced.

CHAPTER 6

WHO'S AFRAID OF NANOTECHNOLOGIES?

In March 2007, France's National Consultative Committee on Ethics (CCNE) delivered its conclusions on nanotechnologies. It expressed concern about the "alarming and ambivalent capacity of molecular man-made nanosystems to pass through biological barriers." A few days later, the headline in the French newspaper *Libération* read: "Are atomically modified objects dangerous to our health?" Nanotechnologies arouse anxiety and have generated an unprecedented wave of discussion, reports and recommendations. Their potential to cause damage provokes numerous questions and sometimes heated opposition. In France, antinanotechnology groups have even sprung up, such as *Pièces et main-d'œuvre* in Grenoble, *Oblomoff* in the Paris region or *Bleue comme une orange* in Toulouse.

So is it fair to say that "the smaller it is, the more evil it is"?[17] Should we be afraid of nanotechnology? It is a complex question, in view of the fact that nanotechnologies encompass objects and manufacturing processes that have nothing in common. Yet, in the eyes of our most severe critics, we nanotechnologists represent a public menace. In 2006, the Grenoble group *Pièces et main-d'œuvre* drew up and circulated a list of French scientists working on nanotechnologies, who were thus to be seen as a threat to the future of humanity. It was a bizarre feeling to find my name on the list. According to this group, my next molecular robots would be capable of passing through biological barriers, penetrating the cells of the human body and modifying their DNA. They would even be able to reproduce. Eventually, they would escape from human control and use all the carbon resources available on Earth to proliferate, destroying the planet and covering it in a vast expanse of gray goo. This possibility was described as the number one risk of nanotechnologies by Bill Joy, scientific director of the computer firm

---

[17]The French satirical weekly *Charlie Hebdo*, November 29, 2006.

Sun Microsystems, who drew worldwide attention to this issue in 2000.[18] However, this is only one of the risks attributed to nanotechnologies.

AMOs: Atomically Modified Organisms

In its report on nanotechnologies issued in March 2007, the CCNE also wrote: "Nanosciences and nanotechnologies aim to achieve human manipulation of the elementary and universal components of matter, atom by atom…." There is nothing impartial about this definition of nanotechnologies. The word "universal" suggests a notion of permanence, immutability. In using it, the CCNE implicitly suggests that the constituents of matter are untouchable, which inevitably leads to the implication that their manipulation is not harmless, that it impinges on the sphere of the forbidden, even the sacred. By using the term "atomically modified objects" instead of nanotechnologies, the newspaper *Libération* enters the same territory, a territory of tacit prohibition, embedded in the deepest reaches of our western culture.

The prohibition on our dealings with matter dates back to the Council of Trent (1545–1563), which confirmed the doctrine of transubstantiation, i.e. the transformation of bread and wine into the flesh and blood of Christ. Since then, the atomistic theory of matter — which is incompatible with this doctrine — has been deemed anathema. Atomism holds that atoms are "specks" of matter, that a speck of bread will always remain a speck of bread and cannot be transformed into a speck of flesh. The Council left its mark on history, and one of those who paid the price was Galileo. The official reason for his condemnation was his support for the ideas of Copernicus, but it could just as easily have been his atomistic beliefs. In the time of St. Thomas Aquinas, it was feared that mice living in the presbytery might nibble the sacramental bread. Today, it is the power of scientists that arouses fear, their capacity to manipulate, divide and assemble matter atom by atom. In a way, nanotechnologies tread on forbidden ground. What will we discover if we play around with single atoms? The secrets of creation? What will we create? Are we playing God, are we not overturning the order of the universe? Will we be able to recreate life? The tip of the STM causes us to slide ineluctably toward questions that are as primordial as they are profound.

---

[18]Bill Joy, "Why the future doesn't need us," *Wired*, April 2000.

Questions on the manipulation of matter inevitably become questions about life. It is hardly surprising that hypothetical "atomically modified organisms" (AMOs) are presented as one of the major threats of nanotechnologies. In the minds of the people who talk about them, they are living organisms modified atom by atom. Following the era of genetic manipulation, the age of the atomic manipulation of life has arrived. The fact is, however, that it is impossible to modify a living organism atom by atom. We would have to know its exact composition down to the smallest atom, then the construction plan, and then be able to insert an ultraminiaturized STM! For it should not be forgotten that at present the only method we have for manipulating matter atom by atom is the STM. These AMOs therefore have no scientific or technical foundation. The term has been invented in order to create an association and imply an equivalence between the technique of atom-by-atom manipulation and GMOs (genetically modified organisms).

At present, scientists can manipulate molecules one by one on the surface of a solid. Soon, they will try to manipulate them on the surface of a cell, then to control the movement of a molecule within a cell. In twenty, fifty or a hundred years, new instruments may perhaps be invented to achieve this. Will scientists ever succeed in assembling and controlling a unimolecular robot able to modify a living organism with atomic precision? Today, this is a question that no one can answer. Should all current scientific exploration be forbidden on the grounds that a future generation of researchers might undertake experiments that are currently impossible? The answer is obviously no. The atom-by-atom manipulation of matter constitutes an instrument for research into the foundations of quantum mechanics and into the boundaries of life, an instrument from which science has much to learn.

Another Threat on the Horizon: Nanomaterials

On January 11, 2007, the authorities in Berkeley, California, decided that research laboratories and industries producing nanoparticles within the city boundaries would be required to fill in a form describing their products and the associated risks. In the same vein, the USA's Environmental Protection Agency (EPA) is planning to regulate the market distribution of consumer goods containing nanoparticles of silver, by requiring companies to provide scientific proof that their product is harmless to the environment. The question arose because of a washing machine that uses such nanoparticles to disinfect items in the wash. The antibacterial properties of silver have long been known, but not the effects of silver nanoparticles on people who might wear the clothes, or on micro-organisms.

The demand for the production and use of nanomaterials to be regulated, in particular those made from nanoparticles, is growing steadily in western countries. Everyone remembers the story of asbestos, which may be responsible for some 100,000 deaths in the next 20 years. Nanoparticles worry people, because it is thought that they might reach the deepest alveoli of the lungs, pass through biological barriers (brain or intestine) and get into the blood more easily than larger particles.

Nanoparticles have been used since the Bronze Age, but up to now the quantity produced has always been very small. The large-scale production of nanoparticle-based materials, without health and safety controls, and their use in the manufacture of consumer goods, could pose grave risks to the public. The nanoparticles emitted by diesel engines (more than 10 million particles with a diameter of less than 100 nanometers per cubic centimeter) or by barbecues are already a matter of public concern. However, many new nanomaterials are currently under development. Let us look at the example of electronic paper. This is a "sheet" on which printed text can be changed at will by means of electric commands. It is composed of electronic ink, made of electrically charged black and white metal nanoparticles, held between two transparent sheets. When an electric field is applied, the black particles migrate to the visible surface of the sheet and display a text. This ink is not a product of nanotechnology, but it contains nanoparticles. Before the paper goes on the market, it would be advisable to study the effect of these nanoparticles, to protect the people who work with them if necessary, and to take measures to prevent their dispersal once the electronic paper is no longer in use.

So nanomaterials made of carbon nanoparticles or nanotubes should be monitored closely. The same is true of many chemical and biochemical products, whether man-made or not. Some 100,000 chemical compounds are in daily use in industrialized countries (in household products, paints, plastics, etc.), and we have information on the toxicity of only around a third of them.[19] Yet not many people are worried about it. Similarly, who

---

[19]Mohamed Larbi Bouguerra, "Ignorance toxique," *Le Monde diplomatique*, June 2002.

is aware of the dangers of refractory ceramic fibers, which have replaced asbestos fibers as a heat insulator in brake linings, catalytic converters or insulation for ovens and boilers? They are classified as "category 2 carcinogens" (substances that present a known risk to animals and should be regarded as presenting a risk to human beings). Glass wool and rock wool, two very common substances, are classified as "category 3 carcinogens" (substances which are of concern to human beings because of possible carcinogenic effects but for which there is insufficient information to reach a satisfactory conclusion). On the other hand, the toxicological data on the carbon fibers found in tennis rackets, boat hulls and bicycles, and on para-aramid fibers (like Kevlar), "are still insufficient for a comprehensive and detailed risk assessment of their effects to be made."[20] There is nothing nanotechnological about these fibers (their diameter is a thousand times larger than a carbon nanocube). They are very widespread, but little is as yet known about their toxicity.

Like the tree that conceals the forest, carbon nanoparticles and nanotubes provide a focus for most public health discussion, and everything else is ignored. Could it be that nanotechnologies provide a scapegoat for an industry that is reluctant to assume its responsibilities for pollution and public health? "It is absolutely essential that producers of new materials should make sure that the safety of their products has been properly assessed, both for workers and for the general public. History shows that this has not often been the case," reads the introduction to the European Union's "Nanosafe 2" program.

Nanotechnologies could provide a model for toxicity and environmental impact studies to be carried out systematically before any mass production. In fact, there are more and more such studies around the world. The European Union is funding the "Nanosafe" program to establish a database on the dangers of nanoparticles. In France, the National Institute for the Industrial Environment and Risks has launched a research program on the risks associated with molecules arising from nanotechnologies. France's National Research and Safety Institute is a center for studies on nanoparticle toxicology. The Environment and Labor Health and Safety Agency looks into the environmental and health risks of nanomaterials. All these studies

---

[20]French Institut national de recherche et de sécurité, special report on fibers, January 2, 2007.

will soon provide insight into the debate. But what will then happen to nano-materials? Some will be banned, others strictly controlled, and the most inoffensive will be mass-produced.

Nanomaterials and nanoparticles are not necessarily synonymous with risk. Homing pigeons possess specialized cells containing magnetic nanoparticles to guide them through the Earth's magnetic field. That is what you might call a harmless nanoparticle! Others are even extremely useful. A spherical particle 5 nanometers in diameter can contain a few thousand atoms, which means that it can be used to deliver a therapeutic substance. As we have said, nanoparticles are capable of crossing the blood–brain barrier which protects the brain (by isolating blood vessels from the cerebral fluid). That is why they entail risks, but also offer huge medical potential. Combined with drugs, they could, for example, be more effective in treating brain tumors, which are currently difficult to treat, as many anticancer drugs do not cross the blood–brain barrier. These drug delivery nanoparticles, which are still at the research stage, are made up of biodegradable polymers or cyclodextrins (molecules containing a cavity) enveloped in lipids. The purpose of the envelope is to give them stealth properties so that they can cross the blood–brain barrier more easily.

Other nanoparticles are being developed to target cells. For example, current anticancer treatments — chemotherapy and radiotherapy — do not distinguish between sick and healthy cells. They attack both kinds of cells indiscriminately and produce significant side effects. Nanoparticles could be used to identify cancer cells. If fitted with a recognition system, for example, they could act as "homing devices," attaching themselves to the diseased cells and releasing their active ingredient. They could also contain a cluster of metal atoms which could be laser-stimulated to heat and destroy diseased cells. Should the production of nanoparticles with such therapeutic effects be banned? Clearly not. But the production plants should be subject to the same health and toxicological precautions as other sectors, such as the pharmaceutical and agrifood industries.

Electronic Spies

Another concern associated with nanotechnologies relates to the sphere of individual liberties. For example, the use of so-called RFID (radio frequency identification) devices is currently spreading very fast. These devices consist of an electronic chip and a radio frequency aerial, allowing the information on the chip to be read and modified remotely, i.e. potentially without our knowing. This is not the case for traditional electronic chips on items such as credit cards, which have to be physically inserted into a reader.

RFID tags, which look like tiny labels or capsules, are already used to provide access to company diners, public transportation systems, or ski lifts in some ski resorts. They are also replacing bar codes to identify goods, works of art, etc. The first cashier-free supermarkets have already appeared, but at the moment customers have to scan their own bar codes. In the near future, supermarket goods will be fitted with an RFID tag, so that instead of placing their purchases on the conveyor belt, all that customers will need to do is wheel their trolleys between radio frequency aerials, which will read the prices.

Some people are worried that not just merchandise, but we ourselves, may one day be fitted with RFID chips. In Belgium, it is already compulsory for pet owners to fit their dogs and cats with an ID chip. Since April 2007, electronic bracelets containing RFID chips have been placed on babies in the maternity ward at Raincy-Montfermeil Hospital in France, as a way to prevent kidnapping. RFID tags could be useful for monitoring elderly patients, for example by notifying caregivers if they fall or wander off. How widespread will they become? Some commentators are afraid that they are paving the way for an era of "traceable human livestock." "The chip, they tell us, enables doctors to intervene more quickly in the event of a problem … Then reasons will be found to 'enchip' ever wider sections of the population. The day will come when we will not be able to live without it. On that day, chips will be automatically implanted at birth. It will become compulsory to carry them. 'Unchipping' will become a criminal offence."[21]

At their current size, RFID tags seem unlikely to threaten our freedoms. However, microelectronics is producing ever-smaller devices, approaching the size of a protein. An alliance between biology and microelectronics has become possible in terms of compatibility of scale, and has already given rise to DNA chips and molecular sensors, microsensors that are sensitive to certain molecules like proteins or RNA, which can be used as detectors.

---

[21] Jean-Michel Truong, in: L. Noualhat, *"Technologie: un grand pas vers la 'traçabilité' humaine"* [technology: a big step towards human "traceability"], *Libération*, May 11, 2002.

Their sensitivity is still low: it often takes thousands of molecules to detect the relevant chemical type reliably. This union of life and technology will continue with nanotechnologies. Some fear that they could transform RFID into nano-RFID, and even into molecular RFID. Nano-RFID tags would become invisible to the naked eye or to optical microscopes, and could act as sentinels to elementary life processes. Inserted in the human body, they would monitor blood composition, free radicals in cells, the release of hormones in the pituitary gland.... Implanted in the brain, they would even be capable of recording signals transmitted by neurons, deciphering nerve impulses without our knowing and transmitting our thoughts. "Several billion minuscule scanners, called nanorobots, could be sent to the brain to record all its details from inside," wrote the American author Ray Kurzweil.[22] A molecule transformed into an RFID device would become the ultimate eavesdropper and a threat to free will.

So the claim is that nanotechnologies are ushering in a world where "there are plenty of eyes at the bottom," to paraphrase Feynman's famous assertion.[23] Invisible nanorobots will be inserted into our bodies, cameras in the form of specks of dust will float above our heads, watching us, walls and ceilings will be coated with powder capable of capturing information and spying on us.... It reminds me of a phone conversation I had with someone who wanted advice, because she thought that her surgeon had implanted nanorobots in her body during a recent operation. She complained that these nanorobots were spying on her and wanted to know if she could come to the lab in Toulouse so that I could remove them!

Is it possible to create so many "eyes" at the molecular scale? Such systems would require an on-board computer, a transmitter and an aerial, all on a single molecule. There is not enough room to fit all that on a molecule. So let us raise the scale a bit and imagine a spy nanorobot that is a dozen nanometers in size. A nanorobot like this is often depicted as a smooth shell fitted with nanograbs to manipulate molecules, which themselves retain their atomic structure. These representations give a false idea of nanorobots, depicting them as macroscopic robots, with macroscopic components, when in fact they will be made up of molecules. All the same,

---

[22]Ray Kurzweil, *The Age of Spiritual Machines* (Texere, New York, London, 2001).
[23]"There's plenty of room at the bottom."

let us suppose that it was possible to build an RFID nanorobot. Once introduced into the human body, it would not be able to stay still and would move around. Its radio frequency signal would then get lost in echoes and would be diffracted around the body. It would be very difficult, if not impossible, to communicate directly with it. In addition, we do not have the slightest idea how we might nanocommunicate information with such a nanorobot. It would be too small for our communication methods. We are capable only of communicating with systems a few micrometers big — in other words, the size of the RFID tags that are being made currently or that will be made in the near future. Not all microelectronic devices can be miniaturized to the extreme: things that are possible on the micrometric scale become impossible on the nanometric scale. There is a great deal of confusion between the two scales, leading to the belief that nanorobots could be fitted with the most sophisticated functions, which is far from true.

But let us put aside the technical questions and suppose that, in twenty, fifty or a hundred years, scientists unexpectedly succeed in manufacturing molecular RFID tags and find a way to exchange information with a single molecule traveling around the human body. What should be done about such nanomachines? Nobody can answer that question, because nobody has the slightest idea what kind of social customs, needs and desires these future people will have.

France's data protection agency, the CNIL, has already set alarm bells ringing on current RFID systems. It considers that RFID tags could contain information which, though apparently insignificant, might say more about the individual than about the tagged object. For this reason, it has brought RFID within the framework of the data protection law covering computer databases.

On the Road to Nanomedicine?

We all want to stay healthy. But how far should treatment go? At a time when "molecular" medicine is a foreseeable possibility, and molecular machines may be able to detect the incidence of a genetic mutation at the molecular scale, some people are asking whether medicine should go that far. Where will it lead? Should we "correct" genetic predispositions to cancers, without knowing whether or not they will declare themselves, or rectify predispositions to minor disorders? Others ask themselves fewer questions and expect this future "nanomedicine" to improve the human body and bestow immortality.

This latter group includes figures like Tom Morrow, Max More and Natasha Vita-More, self-proclaimed futurologists, who head the WTA (World Transhumanist Association). The "transhumanists" exalt the virtues of progress, and promulgate "the moral right to use technology to extend their mental, physical (including reproductive) capacities." "We wish to fulfill our potential by transcending our current biological limitations," declare these individuals, who seek to change the course of human evolution to bring about the emergence of new human beings, whose precursors they claim to be.

What have nanotechnologies to do with this? Max More answers with an example taken from the field of nanomaterials: "We could weave molecules of diamond into the bones of our skull in order to become practically indestructible. At present, our brains are very vulnerable, whereas with diamond nanofibers, a truck driving over your head would cause only minor discomfort."[24]

More than anything else, the transhumanists want to stay young, not only for longer, but for ever. A dream as old as time, dating back at least to 4500 BC, when the early Chinese alchemists were already seeking the key to immortality.[25] They did not find it, and no one has since. The transhumanists place their hopes in genetics and eagerly follow experiments in genetic manipulation, in the expectation of seeing GMI, genetically modified immortality. They also have high hopes for artificial organs, which will replace human organs like spare parts for a car. They imagine themselves as cyborgs, half-machine, half-human, placing implants in their heads containing whole encyclopedias (ideal for memory lapses) or artificial eyes with zoom and infrared options (practical for night vision). "We already wear contact lenses and are fitted with artificial hips. Why not increase our capacities?" they ask. Stelarc, an Australian transhumanist artist who has had a third Teflon ear implanted (though on his arm, which somewhat undermines its functionality), claims: "One day, we will all be wanting

---

[24]Transfert.net, August 1, 2001.
[25]Serge Hutin, *L'Alchimie*, Paris, PUF, coll. *"Que sais-je?"* 2005.

implants to increase our knowledge, our intelligence. I am just a little ahead of you."[26]

The transhumanists feel dissatisfied with this body, which constantly ages and finally — the ultimate insult — dies. Since this body is a problem, why not get rid of it? Live without a body? We could "simply" scan the contents of the brain, download them into a computer memory and be resuscitated at will, either in another body or, for the real extremists, in the form of software in virtual space, a "digital paradise."

How about that for a thrilling prospect? "In the USA, we have the Amish, a religious group which for years has restricted the use of certain technologies in order to maintain its social equilibrium. To draw a parallel, there will undoubtedly be "Humanish" who will choose to remain strictly human, not to use genetic manipulation, not to increase their intelligence, not to live longer," explains More, with a touch of condescension toward these lost sheep.[27]

When asked "How far should treatment go?," the transhumanists set no limits. Their sympathizers include well-known researchers:[28] Marvin Minsky, the father of artificial intelligence, Hans Moravec, the "pope" of robotics, and Ray Kurzweil, the inventor of the synthesizer, who give them scientific credibility. Kurzweil believes that human beings will be immortal within twenty years, thanks to millions of microrobots the size of a blood cell, which will repair our organs night and day, creating new tissue.[29]

Paradoxically, the people who are keenest to highlight the ideas of the transhumanists are their opponents, who use them as a scaremongering tactic: "Look what kind of world we can expect if we allow advances in nanotechnologies, genetics, information technology," they say. They attribute considerable impact to this movement. Yet, in September 2007, the transhumanist association had 4519 members, half in the USA and a third in Europe — not exactly an army.

---

[26] www.stelarc.com

[27] Transfert.net, August 1, 2001.

[28] Rémi Sussan, *Les Utopies posthumaines*, Sophia-Antipolis, Omniscience, coll. *"Les Essais,"* diffusion PUF, 2005.

[29] Ray Kurzweil and Terry Grossman, *Fantastic Voyage: Live Long Enough to Live Forever* (Rodale, 2004).

Potential Military Applications

In the Disney cartoon *The Magic Sword: Quest for Camelot*, the wizard Merlin and the witch hold a contest of magic. As a last resort, Merlin deploys his ultimate weapon — he changes into a virus and becomes invisible. The opponents of nanotechnologies object to their possible military applications, for they believe that an invisible weapon is more dangerous than a visible weapon. Public research laboratories have attached the magical prefix "nano" to their work to receive funding for the purpose of "perfecting" weapons systems. In reality, however, most of them simply use traditional miniaturization techniques.

There is no specific role for nanotechnologies in military technology. Thus, when the Massachusetts Institute of Technology received substantial funding from the US military to create its Nano-Soldier Institute, it was essentially jumping at a great opportunity to collect the financial manna being doled out to nanotechnologies. At MIT, they are studying how nanomaterials can improve battledress to treat a wounded soldier or increase his or her chances of survival in a hostile environment. But they are not inventing invisible weapons, whatever the antimilitarists think. Nor are they designing a secret nanobomb. In fact, what would a nanobomb be like? What military advantages would it bestow?

Whether or not it is possible to manufacture nanobombs, whether they are more dangerous than minibombs or macrobombs, the problem lies not in weapons technology but in the exploitation of new scientific developments for military purposes. There is nothing new about this debate, and we know the answer: until the world is totally and permanently demilitarized, countries will continue to use the best work of their scientists to maintain national security. This is a societal issue between militarists and antimilitarists. The issue is the same, whatever the size of the engine of destruction.

Where Next?

In recent decades, many new technologies have emerged. "Too many!" some say. "Too many for our civilization and too many for our planet." The German sociologist Ulrich Beck has studied the origins of this feeling: we have shifted from an industrial society in which progress benefited the

majority and in which risks, such as workplace accidents, were circum-
scribed, to a society in which the risks have become planetary and of potentially colossal magnitude (pollution, nuclear accident, etc.). The risks have
become "untamable," even by governments. Governments are accused of
no longer protecting their citizens and even suspected of concealing information (contaminated blood, the impact of Chernobyl, etc.). Progress ain't
what it used to be.

The purported danger of nanotechnologies has become a political issue.
In the UK, Prince Charles has warned of the "enormous environmental and
social risks" associated with nanotechnologies, leading Science Minister
Lord Sainsbury to commission a report from the Royal Society and the
Royal Academy of Engineering.[30] In France, nanotechnologies were discussed at a hearing on the topic of "potential risks, ethical issues" in the
National Assembly in November 2006. A member representing the Isère
region proposed that "France should take the initiative, at international level,
of creating a body to monitor and oversee nanotechnologies," whereas the
ecology party members from Grenoble asked for a moratorium on research
into the nanotechnologies in their city.

"Too many!" say antitechnology militants too, demanding that the precautionary principle should be strictly applied to nanotechnologies and that
research should be stopped. This rejection is accompanied by a refusal to go
on consuming. "We say no to electronic gadgets. We refuse to go on buying
more and more useless gadgets, that are polluting to make and to dispose
of." "No to nanofashion junk!" shout the militants of the *Pièces et main-
d'œuvre* group in France. These antinanotechnology claims are linked with
current demands for an end to economic growth, on the basis that ending
economic growth requires the abandonment of scientific research. Because
scientists are supposed to be the source of the production of knowledge,
where all these polluting new technologies originate, they believe that there
is no other solution than to shut down the research labs: "More and more of
us are refusing to pursue economic development and research, the hollow
watchwords of a futureless future."[31] But could a civilization like ours, in
the countries of the West, simply stand still? Is it not, like our planet, a

---

[30]*Nanoscience and Nanotechnologies: Opportunities and Uncertainties*, report of the Royal
Society and the Royal Academy of Engineering, July 29, 2004.
[31]From a tract published by the Oblomoff Committee on March 20, 2007.

system in constant evolution? Is it possible to prevent human beings from innovating?

What do the countries of the southern hemisphere think of these attitudes of rejection of progress and our mistrust of nanotechnologies? The fact is that this distrust is specific to Europe (where one is hard put to find a positive word about nanotechnologies, outside the scientific community) and to North America (where it is less marked, since a majority of the population is interested in scientific and technical advances). In Asia and in the developing countries, the attitude to nanotechnologies is positive, because they are seen to bring hope: hope of a new scientific adventure and hope of economic development, from which these countries will not be excluded. For example, in Singapore, a country virtually devoid of raw materials, the invention of electronic technologies that used very few raw materials would be a tremendous windfall in reducing imports and increasing economic development. It is a country that is doing much to support research into nanotechnologies.

Big countries like China or India are following the western path, i.e. the route marked by America's powerful NNI (National Nanotechnology Initiative). Nonetheless, a few young scientists in these countries, or in other developing countries, are keen to grasp the opportunity that nanotechnology presents and to start a new scientific adventure off the trail trodden by the NNI. The resources of the planet are not infinite, so they are looking to invent a form of development that is not, like ours, based on nonrenewable and scarcely recyclable materials, but which uses as few raw materials as possible. They want to show the countries of the northern hemisphere that it is possible to develop whilst preserving the planet.

In Search of Common Sense

In the debates on nanotechnologies, people discuss science fiction fantasies as if they were reality, examine the apocalyptic scenario of gray goo with absolute seriousness, and think up new acronyms like AMO, which refer to nothing real. Isn't that what we should be worrying about? As long as people spend time on questions like these, they ignore issues that are real and much more important. As long as the focus remains on these topics, it is diverted from matters of much greater urgency.

There is no doubt that scientific progress raises questions. Should we continue research when Murphy's law, which says that anything that can go wrong will go wrong, has never been disproved? In the name of the precautionary principle, in the name of an unknown future, some people are calling for a moratorium on nanotechnologies. They worry: will we be able to control what we create? The philosopher Paul Virilio wrote: "To innovate the ship means to innovate the shipwreck; to invent the steam engine, the locomotive, meant to invent the derailment, the railway disaster ... every period of technological development brings, with its batch of instruments and machines, the emergence of specific accidents, the 'negative' image of the rise of scientific thought."[32]

The Luddites ask us to stop there, not to move on with nanotechnology, or at least to take a break. But nanotechnologies are not the problem. As in every other era, what is at stake here is the very nature of the spark that drives individuals to seek to understand the world around or within them. This spark, like an essential little daemon or a good fairy, is hidden in each of us. No one has the power to extinguish it.

---

[32] *Un paysage d'événements*, Paris, Galilée, coll. *"L'Espace critique,"* 1997.

APPENDIX I

A SHORT HISTORY OF MICROSCOPY

If you want to look at a molecule, there is no point using a magnifying glass. A molecule of water measures 0.3 nanometers, i.e. 0.3 millionths of a millimeter. A benzene molecule is slightly larger, at 0.5 nanometers. Carbohydrates and lipids, composed of carbon, oxygen and hydrogen, are big molecules up to 1 nanometer in size. Proteins, which are composed of amino acids, are 10 times larger, at around 10 nanometers. DNA is a giant molecule, a "macromolecule," which can be several molecules in length, and even several meters when unfolded. But even a giant molecule like this cannot be seen with a magnifying glass.

What instrument can we use to see a molecule? An optical microscope, working with light, allows us to see an enlarged image of a minuscule object in the lens. Nowadays, you can go down to the superstore and buy a microscope capable of enlarging an object a thousand times. For a hundred dollars or so, you can observe the hairs on the legs of a flea, animalcules swimming in a drop of water or the facets of a fly's eye, as did the first microscope users, Robert Hooke, Antonie Van Leeuwenhoek and Galileo, in the 17th century. With this kind of microscope, a spider's web or a hair 50 micrometers across will have an apparent diameter of 5 centimeters. In fact, a water molecule should have an apparent size of 0.3 micrometers and be visible! Unfortunately, despite the efforts of microscopists, no image of a water molecule has ever appeared in the lens of an optical microscope, nor will it ever do so. That is because of light: it breaks up ("diffracts") as it reflects off the object, when that object is of the same order of magnitude as the wavelength of light, which happens to be around one tenth of a micrometer. So the image appears blurred or disappears from the lens.

Other microscopes use light that is invisible to the naked eye (ultraviolet, infrared, etc.) or beams of particles, or even quantum phenomena, like the STM, which has played such a big part in this book.

Nowadays, microscopes are divided into two categories. Either the light or particle source is situated a long way from the object, in which case the device is called a far-field microscope, or it is very close, in which case the device is called a near-field microscope (the distance being measured in relation to the wavelength employed). To understand this difference, let us look at the good old magnifying glass used back in 1668 by the Dutchman Van Leeuwenhoek. A magnifying glass is in fact an optical microscope with a single, almost-spherical lens. Light, either from the Sun or from a lamp, is captured by a set of mirrors and reflected on to the object. It then passes through the lens and shows an enlarged image of the object, which we can see directly. Because the light source is distant, this device is classified as a far-field microscope.

A blind person can use touch to "see" an object: by running a fingernail over it line by line, they can form a mental picture of the object. This method is a version of the "near-field" technique. When the fingernail is replaced by an extremely fine needle, the scanning process becomes more accurate and a very well-defined image can be reconstructed. Near-field microscopy was invented to replace far-field microscopy in situations where the latter is blind because of the phenomenon of diffraction.

In order to gain a better understanding of the main developments in the history of microscopy, we will take a step-by-step journey through the adventures of a molecule that for almost a century has benefited from all the advances in microscopy (and other techniques): copper phthalocyanine (Figure A.1). This is a medium-sized molecule, which has to be placed in a solution, crystallized or deposited on the surface of a solid, in order to be observed. It was synthesized for the first time in 1927. It is in very widespread use today, since it is the molecule that gives plastic garbage sacks their blue coloration.

X-Ray Diffraction

The story begins in 1933, when the English chemist Patrick Linstead embarked on the study of the atomic structure of copper phthalocyanine. The technique he used was x-ray diffraction, which is not microscopy, since it yields an "indirect" picture of the molecule. As its name suggests, it is based on the phenomenon of diffraction, the very thing that causes such

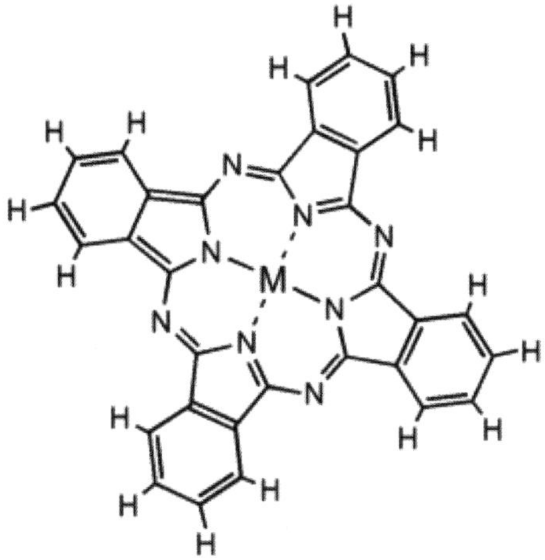

Fig. A.1. Chemical structure of the molecule of metal phthalocyanine synthesized for the first time by the Swiss chemists H. de Diesbach and E. von der Weid in 1927. With copper phthalocyanine, chemical formula $C_{32}H_{16}CuN_8$, the central atom M is a copper atom. This atomic structure was determined for the first time by the technique of x-ray diffraction in 1936. Giving the overall size of this molecule, the imaginary diagonal that passes through M and two of the central nitrogen atoms N measures 1.5 nanometers.

problems in microscopy. X-rays directed at a crystal of the analytical sample provide a geometric image reflecting the interatomic distances, from which the crystal's structure can be reconstructed.

In a crystal of molecules, billions of identical molecules are stacked up. Held in place by their neighbors, they move little, which is crucial for generating an image. When the crystal is thin enough, visible light can pass through it. However, the wavelength of this light (between 400 and 800 nanometers) is much too large to obtain information, somewhat like trying to pick up a nut with a bulldozer. Much shorter wavelengths are needed, and x-rays, with a wavelength equivalent to the distances between the atoms in a crystal, a few tenths of a nanometer, are perfect. It is because of x-rays, to cite just one of the first examples, that we know the structure of sodium chloride, or table salt: a mesh of cubes with edges measuring 0.4 nanometers, topped with chlorine and sodium ions.

Linstead handed over his copper phthalocyanine crystals to a young researcher called John Robertson, who after very extensive calculations managed to establish the organization of the molecules in the crystal

(Figure A.1) and to understand the structure of the molecule itself — a square with sides measuring 1.3 nanometers.

Copper Phthalocyanine in Pictures

Today's large electron microscopes work on the same principle. A very fine metal needle is positioned opposite a metal plate. If sufficient voltage is applied between these two components, the needle starts to emit electrons. Everything then depends on the distance between the tip and the plate.

The "field emission" and the "electron" microscope were invented in the 1930s. A long and bitter competition immediately ensued between the teams working with these machines. They were competing to be the first to achieve atomic resolution, i.e. to obtain a picture of an atom on a screen. In a speech to the academies of science in the late 1960s, Gaston Dupouy, one of the pioneers of electron microscopy in France, explained: "My goal is to see atoms themselves." The race began in Germany in the Telefunken (now Siemens) laboratories. The two young men leading it were called Erwin Müller and Ernst Ruska.

Erwin Müller was a curious and stubborn individual. He was absolutely determined to demonstrate that the beam of electrons emitted by the tip contains information on the position of the atoms of that tip. By positioning the phosphorescent screen sufficiently far away, he thought that he would be able to produce an enlarged projection of the arrangement of the atoms, through a kind of Chinese puppet show effect. To test this idea, in 1936 he invented the field emission microscope. Unfortunately, try as he might, he failed to produce a picture of atoms. In 1951, however, an experiment went wrong, and a small quantity of hydrogen accidentally contaminated the container around his device. Müller accidentally reversed the polarity of the voltage between the needle and the plate, thereby creating hydrogen ions on the surface of the tip. Projected instead of the electrons, they outlined on the fluorescent screen … a picture of rows of atoms!

He continued his experiments with heavier, relatively inert gases like helium and neon, preparing the tip of the tungsten needle with ever-greater care, until he saw the tungsten atoms present on the surface of the needle displayed on the screen. Erwin Müller is thus the first man to have obtained a picture of a single atom. This was in 1955. It was not until 1991 that

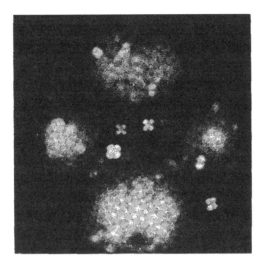

Fig. A.2. Field emission microscope image of a few molecules of copper phthalocyanine deposited on a tungsten needle. It was recorded by E. Müller in his lab at the University of Pennsylvania in 1957. Each molecule is shaped like a small cross with four branches, clearly identifiable in the picture. This picture is a reproduction of the photograph of the phosphorescent screen used by Müller, showing a projection of the electrons emitted by the tungsten tip, some of which are passing through molecules.

Don Eigler, working at IBM in California, became the first man to touch an atom with the tip of an STM. But that is another story, told in Chapter 3 of this book.

The sequel to Müller's experiment is less well known. Deliberately or accidentally, he introduced a small amount of copper phthalocyanine into his container. A few molecules were deposited on the tip. Projected on the screen were images of tungsten atoms, as before, but also strange little clouds with four symmetrical lobes. The distance between two tungsten atoms at the end of the needle was known, and was used as a reference to measure the dimensions of the lobes. He compared them with the distances obtained by Robertson with x-ray diffraction, and the dimensions matched. In 1957, using field emission microscopy, Müller had obtained the first image of a single molecule (Figure A.2). Once again, copper phthalocyanine was at the center of a scientific first.

Müller would produce many more images of other molecules. At this time, the field emission microscope had a fifteen-year lead over its competitor, the electron microscope. However, it was a difficult technique and required extreme electric field and pressure conditions. Moreover, unlike x-ray devices and the electron microscope, it did not reveal the atomic structure of the molecules. These days, the field ion microscope is used exclusively for characterizing tip structure at the atomic scale — essential information for STMs.

The Birth of Electron Microscopy

In the early 1930s, Ernst Ruska, an engineer at the University of Berlin, was given the task of studying the parameters required to control the diameter of the spot formed by a beam of electrons traveling through a drilled metal plate. He first confirmed that a solenoid (a spool of electric wire acting as a magnet) causes a variation in the diameter of a beam of electrons passing through it, in the same way as a convex lens acts on a beam of light. He took the analogy between electrons and light further, and built a transmission microscope with a source of electrons, a solenoid and a projection screen (all, of course, in a vacuum). He placed the small object to be magnified between the solenoid and the screen, with the solenoid playing the role of the magnifying lens in an optical microscope. He achieved a magnification of 14.4 — he had just invented the electron microscope.

Knowing the theoretical limits of light-based imaging, Ruska then had a hunch that his electrons could provide better resolution. Alas, Louis de Broglie's 1927 thesis dashed his hopes, since it demonstrated that the electron has a wave associated with it, like any material particle. Ruska had hoped to evade the theoretical limits of optics, only to find that his microscope would also be subject to wave physics, and therefore to diffraction. However, he did not lose heart, and in 1932 demonstrated that the resolution limit of the electron microscope would be 0.22 nanometers. Here was a source of new hopes — it was theoretically possible to see atoms!

The race for performance was now on. At the end of the 1930s, the electron microscope could magnify 30,000 times, a figure that had risen to 100,000 times by the 1950s. To see an atom, the magnification would have to rise by a further factor of 1000. All these advances took place at Telefunken, a company full of young researchers totally committed to the challenge, who contributed in parallel to the development of television. Everything

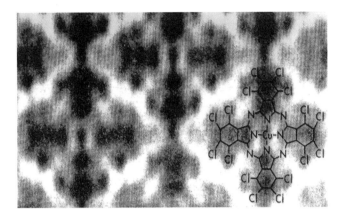

Fig. A.3. Transmission electron microscope image of an ultrathin crystal of copper phthalocyanine obtained by H. Hashimoto in his lab at Tokyo University in 1974. The chemical structure of one of the molecules is superimposed on the image recorded on an ordinary photographic plate after an electron beam has passed through the crystal.

was tried: transmitting electrons through the sample, reflection, scanning with a fine electron brush and combinations of these strategies ... but the atom was nowhere to be seen. It was only in 1970 that electron microscopy finally revealed the first atoms, not in Germany but in the USA, by means of a combined scanning and transmission electron microscope.

In 1974, our copper phthalocyanine molecule was once again in the news. A researcher at Tokyo University chose it as his study model, since it has a copper atom in its center, which was visible with a transmission electron microscope. Even if the microscope could not show all the carbon and nitrogen atoms in the molecule, H. Hashimoto hoped to see a regular lattice made up of copper atoms. So he placed a small crystal of copper phthalocyanine in his microscope and obtained superb pictures, more detailed than in field ion microscopy (Figure A.3)! The technique had another advantage: it did not require extreme pressure conditions, and the image was directly observable, in contrast to x-ray diffraction.

The Scanning Tunneling Microscope

In the microscopes we have just described, the emitting tip is a very long way from the observation screen. What happens when the needle is brought closer to the metal screen? The needle and the screen form a minuscule capacitor, which can be given a charge by applying a voltage. For example, the application of around 1 volt causes a few electrons to accumulate on the capacitor terminals (when a distance of a few nanometers is maintained between needle and surface). Because the polarization voltage (the voltage at the capacitor terminals) is low, the electrons are not emitted by the tip, in contrast to the electron microscope.

However, this tiny capacitor has one defect: it is so small that its two armatures interact electronically across the space between the tip of the needle and the surface. This means that an electron "is not quite sure" which armature it belongs to. This quantum "hesitation" produces a small leakage current, called tunneling current. With a polarization voltage of 1 volt and a distance of 1 nanometer between the needle and the surface, its intensity is in the region of 1 nanoamp and decreases as the needle moves away from the surface. This tunneling current may seem weak, but it nevertheless corresponds to a transfer of 10^{10} electrons per second between the tip and the surface. This weakness was to become a major strength in the hands of Heinrich Rohrer and Gerd Binnig, two researchers at IBM's research laboratory in Zürich in Switzerland.

In the late 1970s, Rohrer was looking at the defects present in ultrathin insulating films deposited on the surface of a metal or semiconductor. These defects were often less than 10 nanometers in size, but they could impair the quality of magnetic storage devices or the reliability of small electronic transistors. At the time, however, there were no microscopy techniques that could examine the structure of these defects without destroying them. Binnig and Rohrer came up with the idea of using the small leakage current arising from the tunneling effect, which could provide an excellent means of describing the electronic properties of a defect and assessing the distance between the tip of the needle and the surface, i.e. establishing the relief of the sample. Along with Christoph Gerber, an IBM research engineer, the two men began putting together an instrument made up of an extremely fine needle, which could be moved toward and away from the silicon surface at will, combined with a system capable of measuring a minute current in the region of a nanoamp. By scanning the surface with the tip and measuring the intensity of the tunneling current, they hoped to be able to reconstruct a line-by-line image, similar to those produced by scanning electron microscopes.

The three researchers took three years to build the first prototype. In 1981, they experimentally verified the variation law for the tunneling current based on the distance between tip and surface. The intensity of this current is divided by 10 when the tip of the needle moves just 0.1 nanometers from the surface. So the movement of the needle had to be extremely precise. They managed to control the tip with unprecedented precision, and to scan the surface of the metal without getting too far away from or too close to it (which might scratch it), using three tiny bars made of a material which changes shape slightly when an electric voltage is applied.

In practice, the tunneling current was kept constant during the scanning, so the needle rose and dipped in line with the surface relief. However, instead of recording regular steps as they expected for the "smooth" surface of their sample, the scanning lines in autumn 1982 revealed a succession of little bumps: the profile recorded by their instrument was giving the precise atomic topography of the surface of the scanned crystal! They had invented the scanning tunneling microscope, a feat that won them the Nobel Prize for Physics in 1986.

The STM is a near-field microscope, since the tip of the needle is held very close to the surface. If the needle accidentally touches the surface, the current reaches intensities of around 100 microamps, i.e. 100,000 times greater than the tunneling current. A shock-absorbing mechanism prevents external mechanical vibration disturbing this regulation. Since its discovery, the STM has been used to observe a large number of metal and semiconductor surfaces, and has resolved as many problems in surface crystallography. A derivative of the STM, the "atomic force" microscope — which uses the forces of interaction between the tip and the scanned surface (van der Waals forces of attraction and forces of repulsion arising from the principle of nonpenetration of atoms) — has since been added to the fleet of near-field microscopes.

Let us go back to our copper phthalocyanine molecule. The needle of the STM scans surfaces and produces pictures of them. What happens if we deposit atoms or molecules on that surface? Provided that these small objects are slightly transparent to the electrons produced by quantum tunneling, could they show up on the image as bumps on the surface? The IBM researcher Jim Gimzewski, working in Zürich in 1987, again chose copper phthalocyanine molecules to check this hypothesis. He deposited a pile of these molecules on one corner of a highly conductive silver surface. He then

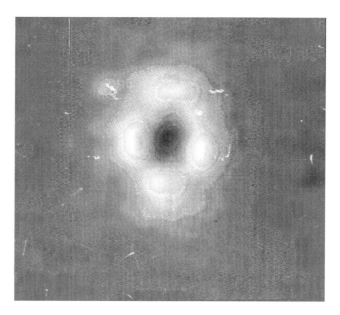

Fig. A.4. STM image of a copper phthalocyanine molecule deposited on the surface of a silver crystal. It was obtained in 1987 by J. K. Gimzewski using his STM at IBM's research laboratory in Zürich. The four lobes correspond to the four lobes of the white crosses obtained using field emission microscopy in 1957 (see Figure A.2). Picture scale: 5 nanometers × 5 nanometers.

dipped the tip of the STM in this pile, captured a few molecules and then raised the tip slightly, so that he could move it to another part of the surface. Here, he brought the tip close to the surface and deposited a small number of molecules. He then cleaned the tip by slightly increasing the polarization voltage. He could now begin the experiment: using this clean tip, he scanned the surface where the few molecules of copper phthalocyanine had been deposited and did indeed obtain an image, the first picture of an isolated molecule produced with an STM (Figure A.4).

This image of a copper phthalocyanine molecule is strangely similar to the one obtained by Erwin Müller thirty years earlier. However, the experiment has been reversed: the molecule is no longer on the tip, but on the surface. An entirely new adventure could now begin, one that was totally inaccessible to field ion microscopy and electron microscopy — for the first time, it was possible to touch the molecule with the tip of the needle, which thus became an extension of the researcher's finger. . . .

APPENDIX II

TRIALS AND TRIBULATIONS OF A PREFIX

According to legend, around 600 BCE the Greek poet Mimnermus of Colophon fell under the spell of a flute player named Nanno. Around the same time, seafarers used to sail from the merchant port of Phocea for the southern coast of Gaul. At this era, the place where Marseille now stands was occupied by a Ligurian tribe that had been colonized by the Greeks. The name of its king was Nann. He had a daughter whom he wished to marry off. A royal banquet was held, at which King Nann's daughter was to choose her future husband. By a combination of circumstances, she chose one of the Greek seafarers instead of her Ligurian suitor. Small, honey-sweetened cakes were served at this banquet — cakes which were still called "nannos" at the port of Marseille centuries later. Our "nanno" sweets were eventually forgotten. Neither the Greek philosophers who "invented" atoms nor — much later — the development of optical microscopy revived the prefix "nanno" as a description of the thousand little objects in nature invisible to the naked eye, even though *nannos* means "dwarf" in ancient Greek. The choice fell on the prefix "micro," from the Greek *mikros*, meaning "small." Associated with the scientific instrument that allows us to see what is invisible to the naked eye — the microscope — this prefix, "micro," came to be the permanent marker of the very small.

The French philosopher Blaise Pascal had no need to borrow from Greek for his ideas about the "animalcules" observed through the first optical microscopes. Later, in his satire *Micromégas*, Voltaire deploys characters who are very large or very small, even atomic in scale. The hero, Micromégas (8 leagues or around 19 miles tall), from the planet Sirius, joins forces with a dwarf from Saturn, a mere 1000 fathoms or 1 mile tall, a sort of intermediate stage between the large and the small. The two companions meet the Terrians, "atoms," who are the main attraction of Voltaire's philosophical tale. These atoms can speak and even do geometry. Voltaire wonders if they have souls.

The first scientific "nanno" did not appear until 1909, in Germany, at a seminar of the German Zoological Society. Hans Lohmann, an eminent professor of zoology at the University of Kiel, suggested naming the microscopic algae that he had observed through his optical microscope *nannoplankton*, on the grounds that *nannos* was the Greek translation of the German word *Zwerg*, which means "dwarf." In the move from noun to prefix, he opted to retain the double "n" in *nannos*. Today, some biologists continue to work in the field of nannobiology and exchange their scientific results in specialist publications such as the *Journal of Nannoplankton Research*. As described in Chapter 5, other nannobiologists are currently working on nannobacteria, bacteria with dimensions significantly less than 100 nanometers.

Lohmann had to create a prefix to describe objects smaller than one micrometer. In the early 20th century, the unit of length used for the size of molecules was the millionth of a millimeter, sometimes called the millimicron or micromillimeter. The study of radiation emitted or transmitted by gases was progressing by leaps and bounds, and x-rays, which have a wavelength a thousand times shorter than that of visible light, had just been discovered. It had become essential to define new submultiples of the meter, as is confirmed by a glance through speeches by Nobel Prize-winning physicists between the 1900s and the 1920s — results are sometimes reported with five zeros after the decimal point. Similarly, in scientific articles of the time, the wavelength of x-rays is estimated in centimeters, i.e. approximately 0.000000001 of a centimeter! Wavelengths are then given in angstroms, after the Swedish physicist Anders Jonas Ångström, one of the founders of spectroscopy. He established the spectrum chart of solar radiation, i.e. its palette of colors, expressing the wavelength of radiation in the form of multiples of a ten-millionth of a millimeter (10^{-10} meters), a unit that became the angstrom (written as Å) in 1905.

Driven by the need for new submultiples of the meter, therefore, the prefix "nano" made its second scientific entrance into history. In the course of the October 1958 session of the International Committee for Weights and Measures, the nanometer was recognized as the billionth part of the meter, following a 1956 proposal by the Soviet G. Bourdoun. The members of the Committee chose "nano" with a single "n" in deference to the rule that Greek prefixes are assigned to multiples and Latin prefixes to submultiples. For example, the prefix for the multiple 1000, "kilo," is taken from the Greek *khilioi* ("thousand"), whereas the prefix for a thousandth (10^{-3}), "milli," comes from the Latin *millesimus* ("thousandth"). In the 1950s, other roots had to be found for new prefixes. Thus, *gigas* and *teras*, respectively "giant" and "monster" in Greek, gave "giga" (10^9) and "tera" (10^{12}). On the same basis, the Latin root for "dwarf," *nanus*, was chosen over the Greek *nannos*, to make the prefix "nano," which refers to the billionth. It should be noted that the rule is not rigidly applied, since "micro" (10^{-6}) is derived from the Greek word *mikros* ("small") and "pico" (10^{-12}) comes from the Italian *piccolo* ("small").

This confusion between the Greek dwarf with a double "n," unearthed by Lohmann, and the Latin dwarf with one "n," espoused by the International Committee for Weights and Measures, has become a source of misunderstandings over the years. In the 1960s, the Consultative Committee for Scientific Language of the French Academy of Sciences was asked to issue a ruling. Georges Deflandre, then Director of the Micropaleontology Laboratory at France's École pratique des hautes études, requested a ruling from the Committee on the double "n" in the term "nannofossil," deliberately constructed in 1959 to be classified within a family of words that already included "nannoplancton," "nannofacies," etc. Gaston Girard, Dean of the Montpellier Faculty of Medicine, was asked to conduct an etymological study of this prefix, and he concluded that both spellings were legitimate. The Consultative Committee finally ruled: the double "n" would depend on the science in which the terms were used. Which is why paleontology and micropaleontology retain the double "n," whereas physics (in particular metrology), medicine and physiology are restricted to the use of a single "n".

This official separation between "nanno" and "nano" did nothing to resolve the ambiguity between the use of the prefix "nanno," meaning "below micron size," and the precise definition of "nano" as one-billionth. In fact, it exacerbated it. It may be that the members of the Consultative Committee failed to realize that progress in physics, technology and chemistry would take physicists to the nano scale, bringing this prefix into the limelight. That is something of a surprise, given the wonderful pictures of atoms and molecules already being produced in the late 1950s. They might also have been alerted by early warnings in the scientific literature and in science fiction stories, reflecting an interest in the infinitely small. At this time, however, people were mainly interested in the infinitely large and in the moon landings, with the manned *Apollo* flights. Paradoxically, the *Apollo* program was possible only because of the miniaturization of electronic components — the famous microchips. But, at the time, it was not yet possible to shrink them to the nano scale. . . .

All the same, by the early 1960s physicists were already able to draw lines 100 nanometers wide on the surface of materials. These advances were still a long way from what goes on in the metabolisms of living organisms. Many physicists are fascinated by these molecular processes and even see them as models for the ultimate machines. So the discovery of the structure of DNA in 1953 led them to the idea that it was possible to store large quantities of information on a few atoms. On their side, molecular biologists were drawing on technology and cybernetics to explain how genes work like a form of machinery. A macromolecule is an extremely small object, invisible to the naked eye and to the optical microscope — so, in a way, it is a nano-object. Yet a macromolecule still contains thousands of atoms, making it much larger than a nanometer. That is a lot for our little "nano" with its single "n." Toward the end of the 1970s, the term "mesoscopic physics" had emerged, as a description of the physics of objects the size of a macromolecule, on the intermediate scale between 10 and a few hundred nanometers. When in 1974 the chemist Ari Aviram, from IBM's research labs near New York, devised a molecular diode, i.e. an electronic component consisting of a single molecule, which would allow current to flow in a single direction, it weighed a lot less than a protein! So, with molecular biology, mesoscopic physics and molecular electronics, research was moving inexorably toward the world of the infinitely small. Nevertheless, our two prefixes "nanno" and "nano" not yet put in an appearnace in the titles of scientific seminars and lectures of the time. They made their comeback in Japan in 1974, which signaled the beginning of the third story of the "nano".

Norio Taniguchi was a specialist in materials science, working on ways to machine materials to nanometric precision. He gave his subject the name "nanotechnology." Over the fifteen years that followed, the new term attracted little attention. It was not until the invention of the STM in 1981 that it gradually came into the limelight. With its ability to move atoms one by one, this apparatus brought the prefix "nano" back into the vocabulary of science. The California-based company Digital Instrument marketed its STM under the name *Nanoscope I*. A new scientific journal called *Nanotechnology* was launched in the UK. Its creator, David Whitehouse, a professor at Warwick University and a specialist in microengineering and precision machining, persuaded an English scientific publisher that research into miniaturization at the nanometric scale was only in its infancy and that the potential market was huge. The first issue was published in July 1990 and in it Whitehouse predicted that the title of the publication would change from *Nanotechnology* to *Picotechnology*, a shift he saw as highly desirable. Having read this book, you will be familiar with the political developments that have since made "nano" a headline issue. Its status as a prefix is extremely practical. It can be attached to any scientific discipline, however venerable, to wrap it in brand-new nanogarments guaranteed to seduce investors. The only discipline that so far has refused to play the nanogame is mathematics. Nano's capture of a portion of the financial resources devoted to scientific research is a global process. In fact, it extends beyond our planet, with claims that fossils of Martian "nannobacteria" have been found on a meteorite from the Red Planet — "nano" has finally stolen the limelight from "nanno," even in biology.

However, in seeking to extend its empire, our prefix "nano" has overreached itself. It is now aiming at new targets, notably in biology, chemistry and mechanics. Under the umbrella of convergence, it is pursuing new horizons: people no longer speak of nanotechnologies, but of NBIC, an abbreviation for "nanotechnologies, biotechnologies, information technologies and cognitive sciences." Diverted from its original meaning of one billionth of a meter, it is attached to anything small and is increasingly associated with microtechnology. The hybrid prefix "micro and nano" is often used to form an extended pressure group and to attract new research funding or capital to build factories. Amidst the confusion, some scientists now prefer to use the picometer, a thousand times smaller, as an unambiguous description of the scale at which they conduct their research. Others refer to atoms to define the technologies they are developing, at the so-called "atomic scale that is an atom technology." And, to complicate things, biology and paleontology preserve a last pocket of resistance for the double "n"....

BIBLIOGRAPHY

Works by Multiple Authors

Observatoire Français Des Techniques Avancées (OFTA)

La Haute Intégration en électronique, Paris, Masson, coll. "Arago," No. 4, 1987.
L'Électronique moléculaire, Paris, Masson, coll. "Arago," No. 7, 1988.
Nanotechnologies et micromachines, Paris, Masson, coll. "Arago," No. 12, 1992.
Microsystèmes, Paris, Masson, coll. "Arago," No. 21, 1999.
Nanocomposants et nanomachines, Paris, Lavoisier, Tec & Doc, coll. "Arago," No. 26, 2001.
Nanomatériaux, Paris, Lavoisier, Tec & Doc, coll. "Arago," No. 27, 2001.

Royal Society and Royal Academy of Engineering

Nanosciences and Nanotechnologies: Opportunities and Uncertainties, London, July 29, 2004.
Lahmani, Marcel, Dupas, Claire and Houdy, Philippe (eds.), *Les Nanosciences, Nanotechnologies et nanophysique*, Paris, Belin, coll. "Échelles," 2004; revised and extended edition, 2006.

115

Other References

Bensaude-Vincent, Bernadette, *Se libérer de la matière? Fantasmes autour des nouvelles technologies* (INRA, Paris, 2004).

Drexler, Eric, *Engines of Creation: The Coming Era of Nanotechnology* (Anchor, 1986).

Gillet, Philippe, *et al.*, "Bacteria in the Tatahouine meteorite: Nanometric-scale life in rocks," *Earth and Planetary Science Letters*, Vol. 175, 2000, pp. 161–167.

Hutin, Serge, *L'Alchimie*, Paris, PUF, coll. "Que sais-je?", 2005.

Joachim, Christian, "To be nano or not to be nano?", *Nature Materials*, Vol. 4, No. 2, 2005, pp. 107–109.

Joy, Bill, "Why the future doesn't need us," *Wired*, April 2000.

Keller, Evelyn Fox, *Expliquer la vie. Modèles, métaphores et machines en biologie du développement*, Paris, Gallimard, coll. "Bibliothèque des sciences humaines," 2005.

Kubbinga, Henk, *L'Histoire du concept de molécule*, 3 vols. (Springer, Paris, 2002).

Kuehr, Ruediger and Williams, Eric, *Computers and the Environment: Understanding and Managing Their Impacts* (report for UNESCO) (Kluwer, Dordrecht, 2003).

Kurzweil, Ray, *The Age of Spiritual Machines* (Texere, New York, London, 2001).

Larbi Bouguerra, Mohamed, "Ignorance toxique," *Le Monde diplomatique*, June 2002.

Leduc, Stéphane, *La Biologie synthétique*, (A. Poinat, Paris, 1912).

Maniloff, Jack, *et al.*, "Nannobacteria: Size limits and evidence," *Science*, Vol. 276, No. 5320, 1997, pp. 1773–1776.

McKay, David S., *et al.*, "Search for past life on Mars: Possible relic biogenic activity in Martian meteorite ALH8400 1," *Science*, Vol. 273, No. 5277, 1996, pp. 924–930.

Monod, Jacques, *Le Hasard et la Nécessité. Essai sur la philosophie naturelle de la biologie moderne* (Éd. du Seuil, Paris, 1970).

Moore, Gordon E., "Cramming more components onto integrated circuits," *Electronics*, April 19, 1965.

Morin, Hervé, "Les nanotechnologies suscitent déjà des inquiétudes," *Le Monde*, April 30, 2004.

Rosnay, Joël de, "Les biotransistors: La microélectronique du XXIe siècle," *La Recherche*, Vol. 12, No. 124, 1981.

Sussan, Rémi, *Les Utopies posthumaines*, Sophia-Antipolis, Omniscience, coll. "Les Essais," diffusion PUF, 2005.

Szybalski, Wacław, in *Gene*, Vol. 4, No. 3, 1978, p. 181.

Toumey, Chris, "Apostolic succession," *Engineering & Science*, Vol. 68, No. 1/2, 2005.

Uwins, Philippa, *et al.*, "Novel nano-organisms from Australian sandstones," *American Mineralogist*, Vol. 83, Nos. 11–12, 1998, pp. 1541–1550.

Virilio, Paul, *Un paysage d'événements*, Paris, Galilée, coll. "L'Espace critique," 1997.